# Where PLUTO Crossed the Path

*Rambles with a Purpose*

*on the Isle of Wight*

John Farthing

Tim Wander

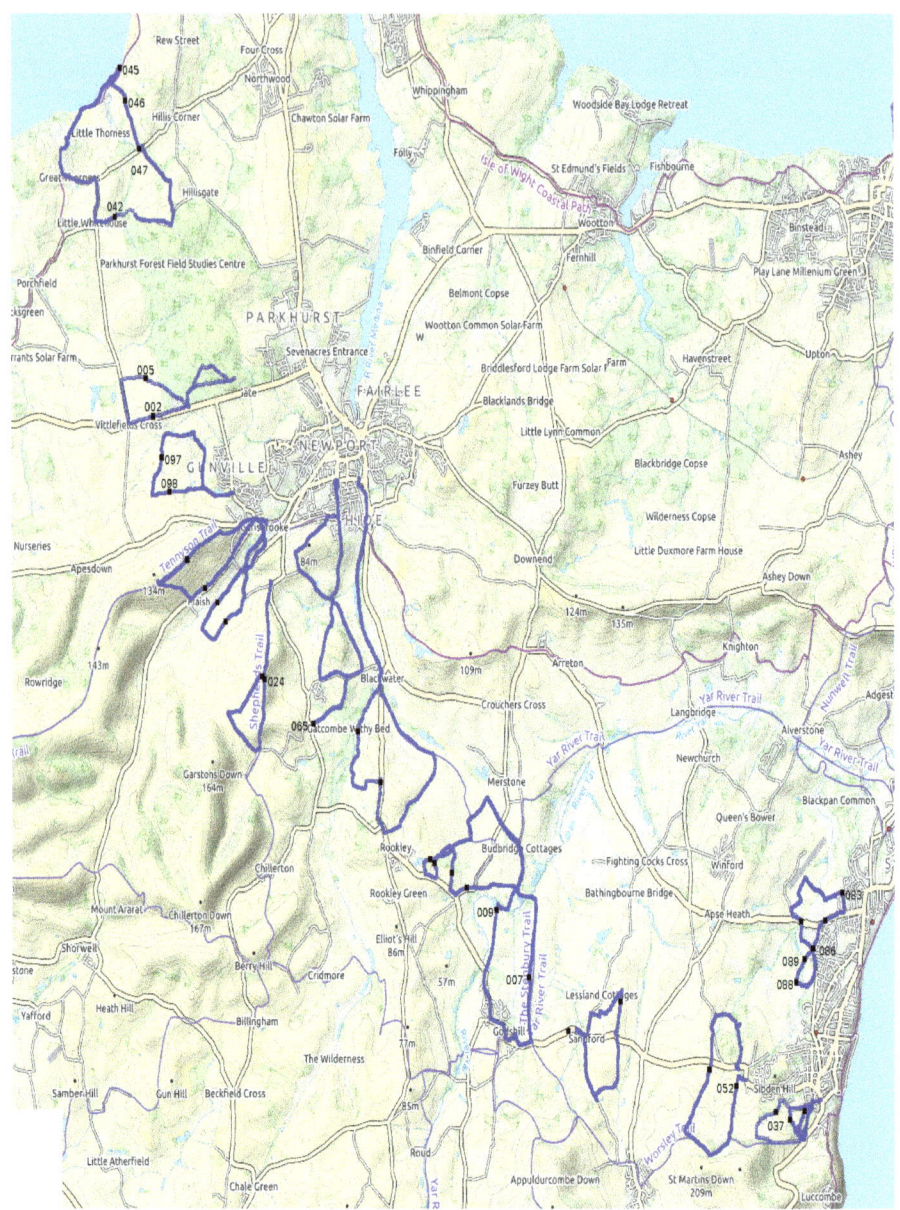

*The routes of all the walks in this book -*

*Walk all these and you will have found Where PLUTO Crossed the Path!*

# Where PLUTO Crossed the Path

## *Rambles with a Purpose*

## *on the Isle of Wight*

First Published by New Generation Publishing in 2019 Text Copyright © 2016 John Farthing and © 2019 Tim Wander. Cover Design © 2019 Tim Wander and Mark Lloyd.

Second Edition

The authors assert their moral right under the Copyright, Designs and Patents Act 1988 to be identified as the authors of this work. A CIP catalogue record for this book is available from the British Library.

ISBN 978-1-78955-750-3

All Rights Reserved. This work is subject to copyright. No part of this publication may be reproduced, stored in a retrieval system or transmitted, in any form or by any means without the prior consent of the author, nor be otherwise circulated in any form of binding or cover other than that which it is published and without a similar condition being imposed on the subsequent purchaser.

Exempted from this legal reservation are brief excerpts in connection with reviews or scholarly analysis or material supplied specifically by the author for the purpose of being entered and executed on a computer system, for exclusive use by the purchaser of the work.

While the information in this book is believed to be true and accurate at the date of publication, neither the author nor the editors nor the publisher can accept any legal responsibility for any errors or omissions that may be made. The publisher makes no warranty, express or implied, with respect to the material contained herein.

Tim Wander can always be contacted via:

www.marconibooks.co.uk

New Generation Publishing

www.newgeneration-publishing.com

# Where PLUTO Crossed the Path

*Rambles with a Purpose*

*on the Isle of Wight*

*John Farthing*

*Tim Wander*

For all the engineers

who gave their all

to allow freedom

to triumph over tyranny.

1939-1945


*The only way to discover the limits of the possible*

*is to go beyond them into the impossible.*

***Arthur C. Clarke***


*Never tell people how to do things.*

*Tell them what to do,*

*and they will surprise you with their ingenuity.*


*My men can eat their belts,*

*but my tanks have gotta have gas!*


*A good solution applied with vigour now*

*is better than a perfect solution applied ten minutes later.*

***General George Smith Patton, Jr.***

There was no ambiguity about the war.

The men who put their lives on the line on D-Day

were prepared to accept sacrifice.

If necessary, the ultimate sacrifice, to protect their countries

and their peoples from a monstrous and evil tyranny.

We owe our peace and freedom today to that readiness on their part.

And for the sacrifice of so many.

We salute them.

They changed our world.

And kept it safe.

From the First Edition

of

# *Where PLUTO Crossed the Path*

The little book was dedicated to

Mr Bill Shepard

*Whose enthusiasm for things*

*historical and botanical*

*should be an inspiration to us all.*

*Bill also 'indirectly' showed me*

*my first PLUTO marker*

*and for*

*Irene who was a great help.*

For Patch.

*Whose last walk was looking*

*for his old friend PLUTO.*

*I hope he finds him.*

All proceeds from the sale of this book, after expenses,
will go to a local charity.

# Where PLUTO Crossed the Path

*Rambles with a Purpose*

*on the Isle of Wight*

### John Farthing    Tim Wander

## *CONTENTS*

PREFACE

AUTHORS COMMENTARY AND NOTES ON ACCESSIBILITY

ACKNOWLEDEGMENTS AND THANKS

PLUTO?

INTRODUCTION TO WALKING THE WALKS

PHOTO CREDITS, COPYRIGHT AND SPELLING NOTES

FOREWORD

PREAMBLE TO FIRST EDITION

POST PREAMBLE

PREAMBLE TO SECOND EDITION

PLUTO ON THE ISLE OF WIGHT TODAY

THE PLUTO MARKERS

THE PLUTO ROUTE ACROSS THE ISLE OF WIGHT

## THE WALKS

1. CRISS-CROSSING SANDFORD
2. TO BLACKWATER AND ON TO ROOKLEY
3. A SHORT SHANKLIN CIRCULAR
4. MARKS CORNER TO THE SOLENT SEA
5. BOWCOMBE VALLEY AND THE LUKELY
6. APSE AND DOWNS
7. NEW PARK AND OLD MANOR
8. MERSTONE STATION TO GODSHILL STATION VIA PLUTO
9. A NEW WOOD AND HUNGERBERRY
10. A FOREST CORNER FROM TUCKERS GATE
11. TO RYDE FORD AND BEYOND
12. UP AND DOWN A DARK LANE
13. ALL AROUND WHITELY BANK
14. THE START OF THE OLD TRAIL
15. A WALK IN CENTRAL FARMLAND
16. BACK 'O' WHITECROFT
17. A SOLO ENTRY ON THE NORTH-WEST SHORE
18. AN EVEN SHORTER SHANKLIN CIRCULAR
19. HILL AND DALE
20. WESTERN LAKE

AUTHORS NOTE - THE ROAD CROSSINGS (JUST TO TIDY UP)

21. SHANKLIN PUMPING STATIONS - A SHORT STROLL ALONG THE SEA FRONT

22. SMALLBROOK AND RYDE - TWO SHORT VISITS TO FIND TOMBOLA

23. FROM THORNESS TO REW STREET - A LITTLE STROLL ALONG THE BEACH

24. A STROLL AROUND BROWN'S GOLF COURSE - AND THE PLUTO PAVILION

## SECTION 2 - THE PLUTO STORY

2.1 D-DAY - 6TH JUNE 1944

2.2 THE NEED FOR FUEL

    2.2.1 THE STORY OF PLUTO

2.3 PLUTO - ACRONYMS AND CODENAMES

2.4 PLUTO - TYPES OF PIPES

2.5. BAMBI - WAS IT A PIPELINE TOO FAR?

    2.5.1 PLUTO - A CONCLUSION

    2.5.2 PLUTO - AFTER THE WAR

## APPENDICES

APPENDIX 1    WALKING ON THE ISLAND IN GENERAL

APPENDIX 2    BIBLIOGRAPHY AND FURTHER READING

    2.2 SELECTED ISLAND BIBLIOGRAPHY

APPENDIX 3    OTHER PLACES OF PLUTO INTEREST ON THE ISLE OF WIGHT

    3.1 MORE PLUTO QUESTIONS TO ANSWER?

ABOUT THE AUTHORS

END-PIECE

# Where PLUTO Crossed the Path

*Rambles with a Purpose*

*on the Isle of Wight*

## PREFACE

For a hundred years the coastline of the Isle of Wight was one of the heaviest and most fortified coastlines in the country. The period 1860-1875 saw a huge array of fortifications, barracks and gun batteries spring up to defend against the risk of a French invasion.

Ninety years later, and for a large part of the Second World War, the Isle of Wight's role was to act as a high density 'gun platform'. Over 30 anti-aircraft batteries together with fixed and mobile searchlight detachments were spread across the island. These were designed to project a wall of light and high explosive steel shrapnel into the air to protect the vital naval ports of Southampton and Portsmouth. Huge gun emplacements, undersea loop-detection, mine loops and radar systems also protected the entrances to the Solent.

Later in the war the Island became a huge garrison camp, refuelling and degaussing site as tens of thousands of troops massed for D-Day. In 1943 the Isle of Wight had been selected to become one of the two sites for the highly ambitious Pipeline Under The Ocean (**PLUTO**) project, designed to provide Allied troops with fuel during the invasion of northern Europe.

For most of the spring of 1944 the Islanders had been fascinated by the peculiar objects which were constantly drifting past the Solent coastline. One day it was an enormous concrete structure that resembled a giant four-poster bed with its feet in the air. Then giant 'cotton' reels started moving out towards the Needles and round the Island to be moored in Sandown Bay. During the summer of 1944 they were to sit there, gathering barnacles, for some four months. The structures were parts for the Mulberry harbour and the giant cotton reels were for **PLUTO**.

Today, almost all of this incredible history has been lost and much of what happened during this time has been forgotten. Also, sadly, nearly all the men and women who served there have also left us.

But the little that has survived is important.

These remains link us to that past.

I think it is important that their story is told and that we remember.

*Tim Wander. July 2019.*

# Where PLUTO Crossed the Path

## *Rambles with a Purpose*

## *on the Isle of Wight*

### AUTHORS COMMENTARY

With any historic research there will always be different perspectives and interpretations. This is only right and should fuel both debate and hopefully further research. There is always more to discover, interpret and understand.

In recent times this process has been complicated by the fascination of 'social media' with conspiracy theories and alternative histories; seemingly sometimes simply for the sake of arguing an opposite perspective, or to gather some dreaded 'likes', whatever they are.

At times this can be helpful; any history needs to be reviewed, updated, researched and questions asked - especially as new information comes to light. The huge coverage of social media and historic websites has become an invaluable tool for researchers and authors, and in some cases, it has allowed whole new histories to be written or clarified.

There have of course also been times in history when 'official narratives' are written to eradicate alternative historical interpretations. It is also true that for various reasons, historical counter-narratives are written that attempt to subvert history, both orally and in print. In the end history is written by everybody, not just, as Churchill once famously wrote, the 'winners'.

Also it is important to remember that any historic research must always adhere to the traditional aphorism that:-

'Absence of evidence is not evidence of absence';

Such that the lack of evidence proving something, does not disapprove it.

In all our PLUTO research, be it with routes and pumping stations, pipeline configuration and power supplies at times we have had to make our best guesses, based on the all the available evidence. We have used our best experience as engineers and historians to try to interpret often conflicting sources (and clear errors that have been continually repeated) and try and work out what actually happened and where, over 75 years ago.

It must be remembered that the PLUTO programme was an enormous and high speed effort that broke new boundaries in engineering and logistics. It was pushed into operation by a nation already stretched to breaking point by four years of war. Of course the same nation, aided by our Allies, was at the same time preparing for the largest amphibious landing in history.

The PLUTO project was rapid, rushed, at times improvised. It was always Top Secret and as such poorly documented, and in place little information has survived.

Through this entire book we have striven to update, and if needed, correct the incredible story of PLUTO on the Isle of Wight as we understand it - and what a story it was.

However, the only thing we are sure of is that there is much more of the PLUTO story to uncover.

## *Notes on Accessibility*

The majority of these walks are across hills and over dales. Although not overly taxing, with proper footwear and appropriate clothing for the sometimes changeable weather, they provide great treks with amazing views all across this lovely Island.

For people who are less mobile, or with less time, several of the walks are far easier, and indeed could be undertaken by wheel chairs.

These are walks 21 around Shanklin, part of 22 to see the PLUTO TOMBOLA pipes at the entrance to Ryde harbour mouth and 24, Brown's Golf Course. The cafe and zoo are easily accessible. The stroll down to the Pavilion does have a path, but this is not metalled. The loop round the golf course is 'off piste', so return to the cafe via the path.

History is not something obscure or unimportant.

History plays a vital role in our everyday lives.

We must learn from our past in order to influence and manage our future.

History serves as a vital model.

It teaches us not only who and what we are to be,

but also allows us to learn what to champion and what to avoid.

Every day, decision making around the world

is constantly based on what came before us.

# Where PLUTO Crossed the Path

## *Rambles with a Purpose*

## *on the Isle of Wight*

### ACKNOWLEDGEMENTS AND THANKS

Firstly, a huge thank you to Helen and Mark Williams who have walked all the walks (sometimes twice!) and carefully vetted, audited and edited all the instructions for this edition of the book. Mark also did a superb job with all the new maps, making it hard to decide which version to eventually use - (although we have retained John's originals for simplicity as well.) They have also taken great photographs and have added significantly to the PLUTO story. Thank you!

As always thank you to my long suffering wife Judith for her typing and putting up with my late night deadlines, and to Mike Plant, a huge thank you for his proof-reading skills, dedication and perseverance. Despite this subject again being outside his usual history of telecommunications' *comfort zone,* as always he somehow manages to keep the text on the straight and narrow or at least makes sure it is almost readable, or at best that it has some semblance of punctuation, capitalisation and sentences that don't need a huge breath midway (like this one:-) Thanks!

We would like also like to acknowledge the help given to John in accessing material by the staff at The County Records Office and The County Archaeological Group and to thank Brian Greening for his comments and advice in getting it into print. ........and finally, Mr Mark Lloyd - yet another great cover!

Of course any and all mistakes, be they literary, grammatical or historical are most firmly all ours.

The most important thing is to enjoy the walks and enjoy the story.

Thank you all for the help and advice.

On to the next one!

*John Farthing and Tim Wander. July 2019*

# PLUTO?

So who or what was PLUTO?

Well there are a few.

In classical mythology, **Pluto** was the ruler of the Greek underworld where souls go after death. He was the brother of Zeus and husband of Persephone. The earlier name for the god was Hades, which became more common as the name of the underworld itself. In ancient Greek religion and mythology, Pluto represents a more positive concept of the god who presides over the afterlife.

**Pluto** (minor planet designation: **134340 Pluto**) is a dwarf planet in the Kuiper belt, a ring of bodies beyond the planet Neptune. It was the first Kuiper belt object to be discovered and is the largest known plutoid (or *ice dwarf*).

**Pluto**, also called **Pluto the Pup**, is the cartoon dog created in 1930 at Walt Disney Productions. He is a yellow-orange coloured, medium-sized, short-haired dog with black ears and is Mickey Mouse's pet. Officially a mixed-breed dog, he made his debut as a bloodhound in the Mickey Mouse cartoon *The Chain Gang*. Together with Mickey Mouse, Minnie Mouse, Donald Duck, Daisy Duck, and Goofy, Pluto is one of the 'Sensational Six' - the biggest stars in the Disney universe. Though all six are non-human animals, Pluto alone is not dressed as a human.

But for our PLUTO story..........Operation Pluto (PipeLines Under The Ocean) was a Second World War operation by British engineers, oil companies, and the British Armed Forces; to construct undersea oil pipelines under the English Channel between England and France in support of Operation Overlord, the Allied invasion of Normandy in June 1944.

According to the Official History, **PLUTO**

originally stood for -

'PipeLine Underwater Transportation of Oil'...

*(but it pumped petrol so they changed it)*

**And our story begins..............**

# Where PLUTO Crossed the Path

*Rambles with a Purpose*

*on the Isle of Wight*

### INTRODUCTION to WALKING THE WALKS!

Following a typically enthusiastic talk by Tim Wander on the PLUTO story a conversation developed which was to lead to the suggestion that we might like to retrace the walks from John's old book *'Where PLUTO Crossed the Path'.*

John's excellent work has been long out of print and as the footways and paths of the Isle of Wight are not static, changes will have occurred. Buildings will have come, and gone and the surviving markers of the now 75-year old pipeline may have been reduced.

In addition, much new information has come to light about the amazing story of PLUTO.

As we have a dog, and like to walk and explore new places, this appealed to us. We started with a few of the walks, and the original hand-drawn maps; followed them, found some updates, and in due course had all the walks written out.

As we had the facility to link GPS tracks into maps we did this as well, initially for our own interest - but it became apparent that it was quite interesting to see how the pipeline had been laid, and where the route led, so the data has been tidied and used, to create the new maps.

In total we covered over well 70 miles of track, along with a fair bit of exploration mileage as well. All of it has been interesting and we have enjoyed tracing the routes and finding the sites and PLUTO traces. It is all about finding the places **Where PLUTO Crossed the Path** and we have been struck by how much remains despite more than 75 years passing and the pipework being ripped out and recycled within two years of D-Day.

It is also daunting, when you walk the route, to understand the size and complexity of the project along with just how fast the project was taken from an idea to a successful conclusion. It is also sobering to understand how quickly it was broken by the needs of warfare, after all the effort to install it.

The original walks were arranged to minimise overlaps and cover the PLUTO route; but also to make interesting walks with a background of the Island's history thrown in for good measure.

This remains the case, and some changes are for the better - if not all - we have tried to alter obsolete references to historical ones, and include some new details. We have enjoyed finding local sources and local knowledge to add and talking to people on the walks.

All in all, it has been interesting and educational - we hope you will get the same satisfaction from trying them all yourself!

Time you all went walking!

*Helen and Mark Williams*

*Helen and Mark; Rambling with a purpose!*

# Where PLUTO Crossed the Path

## *Rambles with a Purpose*

## *on the Isle of Wight*

PHOTO CREDITS AND COPYRIGHT NOTES

All photographs are © Authors Collection, except page 21, 195, 196, 206, 297, 298, 299 and all the coloured maps reproduced by kind permission of Helen and Mark Williams. Photographs 201, 216, 227, 233, 247, 251, 257, 264. © Imperial War Museum. 1945 aerial photographs from Google Earth. Images on page 191, 236 (top), 252 (top) from iwhistory.org.uk, although they are also in the National Archive. Image 236 from Pathe News. Colour maps from www:openstreetmap.org, annotated by Mark Williams.

Every effort has been made to fulfil requirements for reproducing copyright material although most of the images are now at least 80 years old, some much older. Many of the images (and other media files) are now in the public domain because their copyright has expired. This applies to U.S. works where the copyright has expired.In Canada copyright is deemed to have expired if it was subject to Crown copyright and was first published more than 50 years ago, or it was not subject to Crown copyright, and it is a photograph that was created prior to 1st January 1949, or the creator died more than 50 years ago. Where a photograph is taken by an employee in the course of employment, the first owner of the copyright is the employer, unless there is an agreement to the contrary. Elsewhere copyright in a photograph is deemed to last for 70 years from the end of the year in which the photographer dies. Obviously this makes it incredibly hard, if not impossible to ascertain true copyright for most photographs taken that may be over 100 years old. This essentially means that even if you own a photograph, unless you actually took it, in law you do not own the copyright. It is also practically impossible (with the exception of certain collections e.g. IWM, to identify the photographer or company that took, and hence *still own* the photograph.)

In 1998, works published in 1922 or earlier were in the public domain, with 1923 works scheduled to expire at the beginning of 1999. But then the US Congress passed the Sonny Bono Copyright Term Extension Act. It added 20 years to the terms of older works, keeping 1923 works locked up until 2019. The name PLUTO in this context clearly relates to the wartime project. The situation of the images of PLUTO (the cartoon dog) is unclear, however, the images of PLUTO in this book are taken from contemporary WW2 cartoons so fall outside the copyright act. If nothing else they are exempt due to 'fair use' and 'transformative use.' Transformative use requires that you change, or transform, the character enough so that it is no longer a mere copy of the original. The resulting transformation is also sometimes called a 'derivative work'.

Please also note that certain photographs may not be protected by copyright. Section 171(3) of the Copyright, Designs and Patents Act 1988 gives court's jurisdiction to refrain from enforcing the copyright which subsists in works on the grounds of public interest. For example, patent diagrams and handbooks are held to be in the public domain, and are not subject to copyright. Part of the aim of this book was to assemble the widest and rarest collection of photographs illustrating the history of the Isle of Wight and Culver Cliff and the surrounding area. If I have inadvertently tripped over an obscure copyright I apologise and will, when proven, immediately update the digital image with the printers giving copyright designation.

## AUTHORS NOTES – SPELLING

For our American cousins who have emailed me about past books, again I do not apologise for using standard English spelling, phrases, terms, syntax and printing rules throughout this text. However, when quoting original sources, we have usually left the original text/spelling in place. Wherever we thought necessary we have also added a separate translation, correction or explanation alongside in parenthesis.

At the request of European readers, I include conversions and translations for now antiquated distance, weight and monetary systems wherever appropriate. *TW*

## PLUTO – OIL or PETROL?

During WW2 all British, American (and German) tanks used petrol engines, in most cases converted aero engines. Jeeps, half-tracks and most trucks were also petrol based as was every aircraft.

Typically diesel engines were heavy, underpowered (for their size), complex and required better tolerances and materials to manufacture. Only the Russian's persevered with diesel engines, but reliability became a huge issue. Indeed, the first T34 tanks were shipped with a complete spare engine and transmission, which was expected to be needed within 100 miles.

Hence PLUTO always pumped petrol, (the American's call it Gasoline) as did the GPSS network designed to fuel British airfields across the country.

# Where PLUTO Crossed the Path

## *Rambles with a Purpose*

## *on the Isle of Wight*

### FOREWORD

Isle of Wight historian John Farthing's original book – *'Where Pluto Crossed the Path'* was first published in 2008 (by Cross Publishing), it offered 18 spectacular walks across the Island. Each walk set out to follow and criss-cross parts of the 14-mile route that the WW2 PLUTO pipeline took as it stretched around Parkhurst Forest and onward across the Island to Shanklin and then round to Sandown. Each of John's walks carefully took the adventurer close to many of the surviving remnants from the original PLUTO project, but never gave their exact location. The fun is always in looking and walking.

During the Second World War it seemed that the entire Isle of Wight was part of a complex, hurriedly installed and integrated defensive system designed to protect from invasion and aerial attack. Defensive forts, gun batteries, emplacements and countless trenches, searchlight batteries, radar stations and anti-aircraft and machine gun posts were reused or built across the Island. Many of these were deliberately mobile – so they could not be mapped by raiders and later targeted.

The threat of invasion made the small Island one of the most intensively defended areas in the country. It also became a testing and proving ground for new technologies and systems to defend it. Radar and radar controlled guns, anti-submarine loop detection systems, hydrophones, remote fire minefields and all manner of anti-aircraft, anti-tank and anti-invasions systems were rapidly installed. The wireless station that would control all shipping for D-Day was positioned high on Culver Cliff, and the PLUTO pipeline below would supply fuel for the invasion beaches.

**All this in one small island.**

For a brief moment in time, these were special places and important, dangerous things happened there.

So in this second edition we have updated and sometimes corrected the original PLUTO walks, added five new ones and provided a new PLUTO story, explanations, some new maps and more photographs.

Thousands of people visit Sandown and Shanklin on the Isle of Wight every year, so we have also provided some very short walks to find PLUTO in those places. These are also suitable for the less mobile walker, but will still bring the reader close to PLUTO, perhaps the most audacious engineering project of WW2.

If this wasn't enough we have then wrapped all the walks with a new and concise history of PLUTO on the Isle of Wight.

The idea was to give walkers something to read while taking a well-earned rest on their travels, or more likely, while waiting for the rain to clear.

*Tim Wander, Cowes, Isle of Wight. July 2019*

# Where PLUTO Crossed the Path

*Rambles with a Purpose*

*on the Isle of Wight*

PREAMBLE TO THE FIRST EDITION

I must admit straight away that my title is not technically correct because the PLUTO pipeline itself never, of course, crossed the Island. Perhaps 'From SOLO to TOTO and beyond' would have been more correct because the pipeline which crossed the Island was the one from SOLO (The pipe(s) under the Solent) to TOTO (The header tank on the other side of the Island). As fewer people will recognise the acronyms SOLO and TOTO but many will know of PLUTO that is what I will call it throughout.

My intention is to give walkers some fun in finding, or spotting, the pipeline markers, or what remains of them, which were placed on its route across the Island where it crossed highways, byways and footpaths; and in doing so to plot its course. Whilst I thoroughly enjoy walking for its own sake it is sometimes nice to have a purpose, if not an excuse.

Apart from actually 'discovering' the course for myself, the biggest problem was to decide in which order to write about the walks without making the detailed route of the pipeline too obvious, thus spoiling some of the fun. This is why the order will appear more haphazard than just going from one side of the Island to the other. The walks can, of course, be done in any order you wish but I chose the first one on the basis that it passes one of the best preserved markers and one that is also the easiest to spot.

The pipeline crossed several main roads which will be dealt with very briefly. At one or two of these crossings the markers are quite clear at others they are obscure, partial, or simply gone! At least I couldn't find them and others have also looked hard.

There are at least thirty-two markers still discernible on, or very near, public rights of way. When I say discernible I have to explain that many are incomplete. However, once the form of a marker has been recognised in detail, those where only some parts remain can be identified.

The reason for the markers was, of course, to ensure that landowners, their workers, and others needing to dig, or otherwise work the land, were aware of the presence of the pipeline and also to mark the route for persons involved in maintenance and troubleshooting on the pipe itself.

The markers had to be concealed from the air, as you will see from their positions. Great care would have been taken to disguise the presence of any pipeline, military or civilian, during the war and the purpose of the markers would have needed to be kept secret due to possible spying activity. The markers were therefore themselves disguised to look like something more familiar! That's not really a stile!

Obviously in the sixty plus years that have elapsed since the end of the war in Europe, many of the markers have been disturbed or destroyed due to the removal of hedges and road widening; others have deteriorated due to the passage of time. Unfortunately, the situation is also likely to have changed since the time when I started walking and searching in the April of 2002, so logically more markers may have disappeared. There is, however, just a possibility that more could be exposed. It is, I'm afraid, a question of 'bon chance'.

Unfortunately, on some of the walks there seem to be no markers at all, e.g. Walk 6; It is not possible to make a complete mystery of this because there are people of a certain age on the Island and some of the older established farming families who know at least part of the route and still recognise the markers. There are others who will have destroyed them quite oblivious of their purpose. So, to those who 'know' please bear with us; it is only intended to be a bit of fun for those who enjoy walking and have an interest in our wartime and some of our earlier history.

There is no doubt that it will be far better to look for markers in the winter when foliage is greatly reduced; even though ivy and evergreens can still be a problem. However, do the walks in all the seasons; you will always see something different. I appreciate that it is not necessary to have a purpose for a walk other than the walk itself but I would like to think that my subject will add a little interest.

Please enjoy your walking.

*John Farthing - Newport, Isle of Wight, 2006*

*Regular patrols along the pipeline were undertaken. Troop and police would walk each section, every day and night, and then double back. However, Isle of Wight residents still remember putting tins under pipeline joints where they crossed over ground, as the joints leaked. A Huntley and Palmers' tin of rare petrol every night during WW2 was very useful!*

*One of the few pictures showing PLUTO being installed on the Isle of Wight.*

*One of the best preserved PLUTO markers on the Island. Can you find it?*

# Where PLUTO Crossed the Path

## *Rambles with a Purpose*

## *on the Isle of Wight*

### POST PREAMBLE

Instructions for the routes of a series of walks can, of necessity, be rather boring and repetitive, 'turn right, turn left, take this or that stile' etc. This is to some extent inevitable so what I have tried to do is to add some interest and variation along the way, hence a bit of history, albeit by no means exhaustive, and a bit of poetry. Whilst it may not suit everyone, I hope that it will not entirely disappoint. However, I feel that I should say a few things about the bulk of my text.

I do hope that poetry lovers will not be too dismayed at my penchant for including snippets of poetry to 'illustrate' a scene which may be far removed from their original context. This is mainly due to being no poet myself. As will be seen I do like poetry which rhymes and is simple; for this reason, most of what comes to mind tends to be of the older variety.

Both in the main text and in the appendices I have voiced my personal opinion on various subjects. If they are not in accord with those of the reader, then we must just agree to differ. My grammar, such as it is, does not agree with that of Bill Gates, but then some of his spellings seem strange!! (I continue to get a red line under 'Domesday'?) Then again, there are many words that he just doesn't seem to know about!

My descriptions of views will seem rather prolonged and overdone to some, especially to Island people who know their Island. Please bear with me on this, as the intention is to include visitors, who, whilst admiring the view, will not necessarily know what they're seeing. The scene setting and historical bits can always be skipped if it is just the route of the walk that is of interest. Due to some of the walks being in close proximity there some inevitable repetitions of points of interest and features in making each walk 'standalone'. The book, although a little larger today, is intended to be carried afield.

*JF, January 2007*

*PLUTO was a massive engineering project. For most of the 14 miles across the Isle of Wight it was buried to avoid detection and for protection against attack.*

# Where PLUTO Crossed the Path

## *Rambles with a Purpose*

## *on the Isle of Wight*

### PREAMBLE TO THE SECOND EDITION

When I wrote my first edition of this book I had not realised that, in addition to the feeder pipeline which ran from Hungerberry Copse down to Shanklin, there was another feeder pipeline running to the old Granite Fort (now the Zoo) northeast of Sandown. This omission was despite it being clearly stated by both Sir Donald Banks in his book 'Flame Over Britain' and by Adrian Searle in his Book 'PLUTO'. There were 'forked lines' from Hungerberry Copse going to both Shanklin and Sandown. I must have been seduced by the greater publicity attached to Shanklin's part of PLUTO!

It was not until Tim Wander and the ARC team started the restoration of the PLUTO pavilion and the part played by Brown's golf course buildings in the PLUTO story that the full import of those words 'forking lines' became apparent to me. Previously I had thought the Sandown pumps had been supplied by pipes running along the seafront. Indeed, it is now clear that not only did the pipeline run around Lake from Hungerberry to Sandown but a further two pipelines stretched across the bay from Shanklin to Sandown, but more of that later!

So, the emphasis has now changed. In my first edition, regarding the path across the Island, I knew of at least forty-two markers and was able to plan some walks taking in many of them. The new walks are more in the nature of a quest to find the route of this spur pipeline to Sandown where, so far, only three marker positions have been identified. They may actually be the only visible indications which remain.

However, Tim Wander has also been successful in establishing the route of the 'Sandown' pipeline in the White Cross and Merrie Gardens area, which includes the most positive marker, but so much development has taken place inland of Shanklin, Lake and Sandown since the Second World War that the remainder of the route remains a mystery and, so far, can only be guessed at.

The three marker locations that have been found so far are included in the few new walks I have described, and where no markers have been found I have made some suggestions but I would ask walkers to search hedgerows and boundaries of paths and roads on a broad line between Hungerberry Copse and the back of Sandown Zoo, wherever public access exists, for the tell-tale concrete marker posts (see illustrations), and hopefully landowners will search where there are no footpaths.

The topography of the land behind Shanklin must have made it particularly difficult to run a gravity fed pipeline from Hungerberry Copse in the direction of Sandown. Despite Hungerberry's slight height advantage it must, surely, have gone one side or the other of Sibden Hill; I will suggest a walk in the area but considerable deviation is recommended, as so far I have not found any trace of markers south of the old Shanklin to Ventnor railway track.

So, good hunting and please let Tim Wander or myself know if you do find any markers or other indications.

*JHVF, January, 2016*

# Where PLUTO Crossed the Path

## *Rambles with a Purpose*

## *on the Isle of Wight*

### PLUTO on the ISLE OF WIGHT TODAY

Due to a very efficient salvage operation in 1945, and the passage of over 75 years, tangible evidence of the PLUTO operation and project is rare. All signs of the huge Hungerberry Wood storage tank above Shanklin or the pipeline itself have long been removed – trust us there is nothing to find in these dense woods apart from a small drain and a series of low banks!

Evidence of the PLUTO pumping stations on Shanklin seafront remain, but you need to know where to look - walk 22 is a must do. Nothing remains at the across-Island pumping station near Whippance Farm but nearby the best preserved PLUTO area is where the pipelines crossed the Solent (codenamed SOLO) and arrived on the beach at Thorness. Here at low tide, at least 12 or 13 of the original 21 pipes are still in situ and can be clearly seen along with a fragment of the manifold on the beach.

At Sandown on Brown's Golf course three buildings that were converted to house PLUTO pumps have survived. These are the Granite Fort, Brown's Ice Cream Factory and Brown's Cafe. Across the Golf Course the fascinating PLUTO power pavilion has also recently been restored.

These sites, if nothing else, have led to a continued interest in the PLUTO project across the Isle of Wight. This is supported by displays at the Shanklin Chine Museum, Sandown Zoo (in the Granite Fort where 14 PLUTO pumps were housed during WW2) and Bembridge Heritage Centre - the latter two today preserving two of only three known surviving PLUTO pumps.

All the walks in this book criss-cross the route that PLUTO took across the Island. In some places there are almost intangible remnants of the pipeline project. As you leave Thorness on the road from Whippance Farm, in one area, the cows' water troughs all sit behind the hedge in a row on PLUTO concrete bases. Even the hedgerow still bends to accommodate the pipeline pressure as it left the Whippance Farm pumping station by the shore. Elsewhere you can still sometimes discern the scars on the roadway where **PLUTO crossed the path,** or find the ceramic pipes that protected PLUTO as it crossed under a road, or dived underneath a gateway into a field.

# THE PLUTO MARKERS.

By far the most iconic remnants to be found today are the PLUTO boundary markers. These were placed in hedgerows to locate the position of the pipeline buried below as it crossed roads, tracks, paths or a major field boundary. Their purpose was to alert anyone digging in the area of the presence of the pipeline and to allow the military police, who walked the pipeline to deter saboteurs or fuel thieves to cross the hedgerow with ease. From a distance the markers could be mistaken for simple fence styles, no doubt a deliberate ploy to avoid detection from the air.

When John first started walking he first found 42 markers and subsequently we have found at least 10 more plus a stack used as ballast and dumped at Thorness and two washed onto the beach at Rew Street. The latter two were in danger of being destroyed by the sea so have been recovered, restored and now have pride of place in Cowes military museum. They sit alongside parts of original PLUTO pipeline and the only known HAIS to HAMEL connection manifold, recovered before it was lost from the beach and now preserved and displayed as a commemoration and celebration of the amazing PLUTO project.

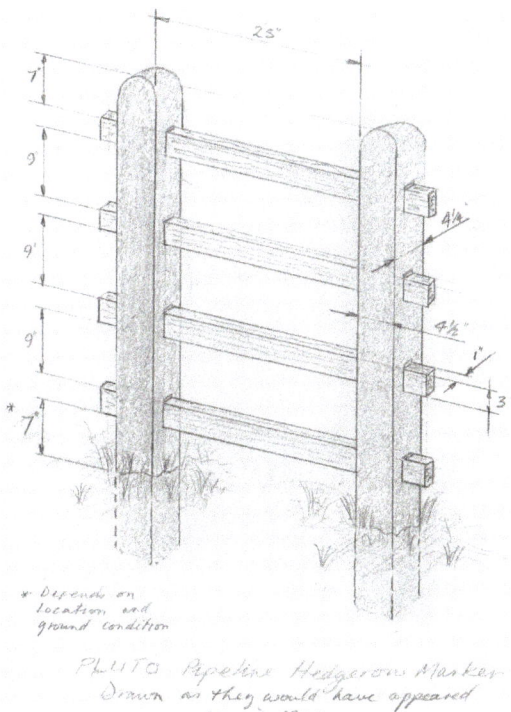

Today many of the PLUTO markers are in a very poor state after 75 plus years while others were in surprisingly good condition. We suspect that at least five of the markers identified in the early 2000s have now been lost. Roads are widened, ditches are redug and hedges are removed – and frankly few people would recognise what the markers actually are or were.

If you would like a clue about what to look for then head down to the remains of the Thorness beach manifold – all the vertical concrete posts are actually PLUTO marker posts. We don't know if they were always there or whether they were fitted later as their slots would make construction of a fenced off area easier. But before you go hunting in hedgerows maybe a walk down to Thorness will give you a clue of what you are looking for.

The idea of the walks is firstly to get people out to enjoy the amazing Isle of Wight countryside. But as you follow the walks the goal often becomes just to find the markers. When you find your first one it is great fun and it will quickly become addictive. The walks give you clues as to their locations and in this edition we have even added some photographs of the actual markers on the route, but with not enough information to actually locate them without some legwork! All we can ask is that you take only photographs and leave them safe for future generations of walkers and historians. Each comprised two concrete posts with rectangular holes to receive up to 4 horizontal oak slats. Today only a few markers retain all their slats and in others often only a single pole survives. Most are hidden in hedgerows, a few have fallen, others have been moved and at least one is part buried.

*Good luck and Good Hunting!*     *JF & TW. July 2019*

# Where PLUTO Crossed the Path

## *Rambles with a Purpose on the Isle of Wight*

THE PLUTO ROUTE ACROSS THE ISLE OF WIGHT.

The PLUTO pipelines that crossed the Solent ran for just over 3 miles underwater. They left the Badminston tanks near the Fawley refinery, ran overland to Lepe, and then across to the Thorness beach Terminal on the Isle of Wight. The main pipeline then ran over 14 miles across the Island into the Hungerberry tank above Shanklin. From there a feeder pipeline looped around the town of Lake, although after Merrie Gardens and the golf course clues to its route into Sandown are then difficult to find. We now think it ran alongside the railway track, before looping into the rear of the granite fort.

The last part of the route into Hungerberry has also been difficult to determine. With much research and having studied the few clues, we have walked the obvious crossing points for railways and roads and studied the quite severe land gradients. In this map we present our best estimation. From Hungerberry the route ran to the head of Shanklin Chine and down the upper Chine is known, but then two theories on its route have been considered. There is strong evidence that the pipeline left the chine near the current public entrance and ran across the park and along Keats Green, before dropping over the cliff before the Shanklin lift. One source even states the pipes were actually hidden inside the empty lift tower. The other route is that it carried on down the Chine to arrive at the beach and then turned east to arrive at the first pump some 250 yards along the beach.

More research is required and large tanks discovered under Rylstone gardens may have had a part to play. There is also the possibility that Shanklin Chine (and indeed Sandown and Shanklin beaches) were provided with a secret anti-invasion beach flame barrage system as early as 1940. Literally fuel would be pumped, or fall under gravity, to flood across the beach, where it would be ignited to prevent invading forces gaining a foothold. Some evidence suggests that a similar system was installed to protect Freshwater Bay, with the fuel tanks known to have been buried in the cliff top above. These could, with the passage of time, obviously get confused with PLUTO in 1944. It is also possible that the PLUTO pipe from Hungerberry split to provide two feeder pipes to supply the spread out Shanklin pumps. This would provide duplication in the event of attack and may have been part of the design to supply the two pipelines that stretched under the bay to supply the Sandown site.

**PLUTO across the Isle of Wight in red.**

The yellow sections show the TOMBOLA refuelling system at Rew Street by Thorness, and our suggested route from Sandown to Ryde.

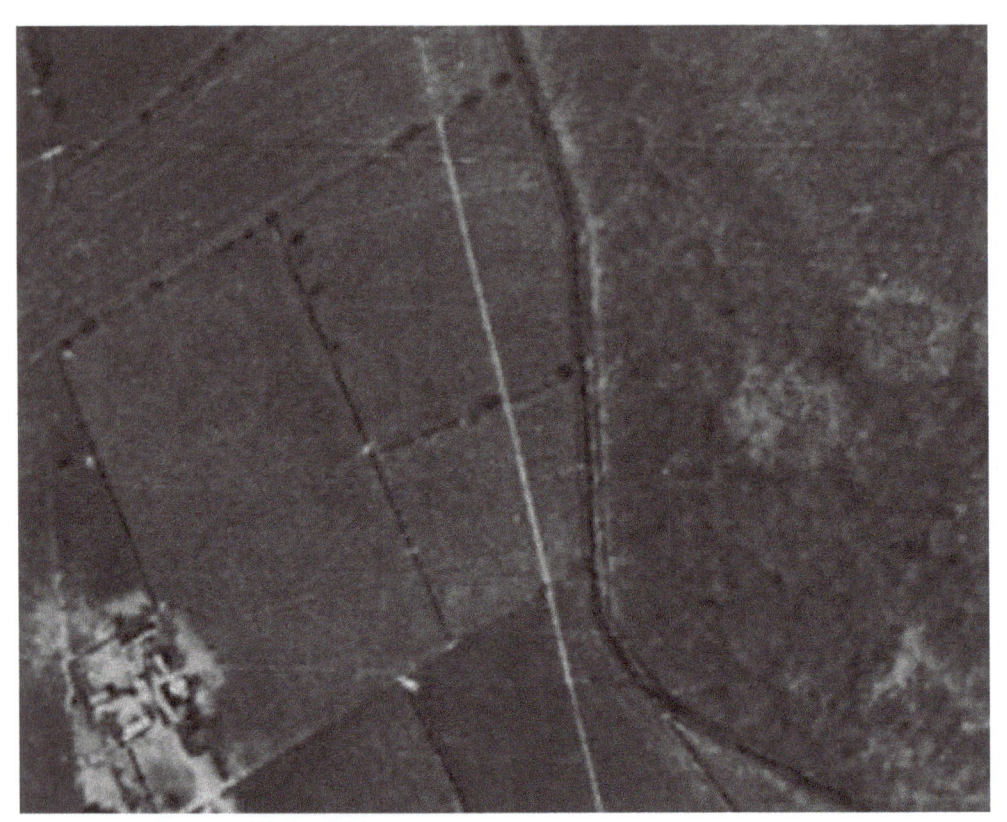

*SO... How do you find PLUTO?*

*Well one method is to find period aerial photographs. Despite all the secrecy, the PLUTO pipeline left a scar across the Island. The above picture was taken in 1946, just west of Marks Corner,*

*In other areas it is less clear. This image taken around 1945 shows the pipeline looping around Lake, near the Merrie Gardens Public House. It is still visible.... if you look carefully - but in other places it disappears. This may be due to it running alongside existing roads, pathways or train tracks. There is also a possibility it was never finished...*

# Where PLUTO Crossed the Path

*Rambles with a Purpose*

*on the Isle of Wight*

# Time to go walking!

## 1. CRISS-CROSSING SANDFORD

*(Including the surrounding farmland)*     *(Just over 2¼ miles.)*

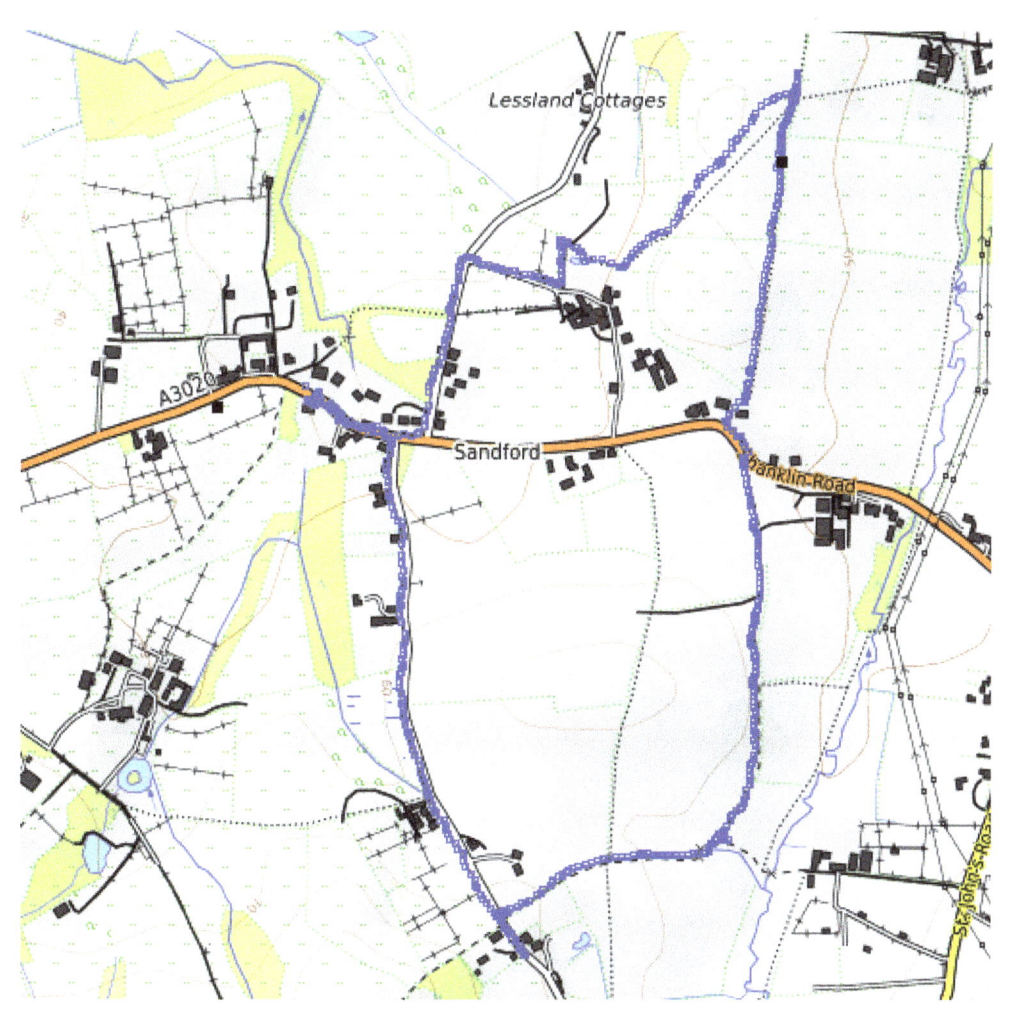

Starting at the little 'lay-by' on the Godshill to Shanklin road outside Sandford House, walk on the verge toward Shanklin on the left-hand side of the road. The hollow wherein a few houses have been built on the opposite side of the road is where the sandy clays were taken from for brick making. After about seventy-five yards, just before Lessland Lane on the left, cross the road (take great care, it's almost blind to traffic from the right) and proceed up Redhill Lane to the right.

This area to the top of the hill further on has for many years been known as Daffodil Valley; there is a house called this near the top of the hill, which was christened first I do not know. There is a scattering of daffodils in the verges; presumably there are more, down in the valley. In the forties and early fifties one could make a small donation at the house and go down and pick them.

This is a typical Island country lane, as yet unspoilt, very narrow and with particularly regular high banks mainly hedged in hazel and often with a very noticeable scent of foxes. At the top of the hill are some sweet chestnut trees; quite productive at the right time of year. Over the bank to the right can be seen, in winter, the little valley itself with glimpses of Godshill Park farm on the farther side.

Over the brow of the hill and down the other side you come first to Park Hill Farm. The footpath GL 43 goes to the right here over to the Godshill Park carriage drive; but this is not for us today. A further rise in the lane goes past Park Wall House on the left. I had understood that this name related to the fact that the wall enclosing Appuldurcombe Park joined the lane at this point; but the mapping of 1863 called this Park Well Cottage! A study of the deeds might resolve this.

Immediately after the end of Park Wall house garden is a bridleway (GL 35) to Sandford on the left; this is the one we require. (Ignore GL 35 which continues on the right to the Freemantle Gate, a magnificent gate in the aforementioned wall, but not for this walk). Island Roads have now resurfaced the lane and the milestone seems to be no more.

Take the bridleway to Sandford, now on the right. This path is fenced both sides initially, and can be muddy in the winter months, but the views to the right are immediately striking. First on the right is Wroxall Down merging unbroken into St. Boniface and then the much closer mass of St. Martins Down; to the front is the high ground rising inland from Shanklin. The gate at the end of this section takes you into open farmland with even better views (and sheep & electric fencing which a dog can easily pass). Ignore the footpath, signed to the left, after about one hundred yards; this is GL 38 and also goes to Sandford but is not on your route. A further hundred yards brings up a compulsory turn to the left over a mini ravine so typical of Island field drains and small streams where the water has cut down through the soft sandstone.

A bridle gate takes you into a small paddock with a good view, to the right, of the old buildings of Lower Winstone Farm; now, a much appreciated Donkey Sanctuary. Although not on this walk, a path from here leads down to it (GL 36).

Continuing the walk. A few steps across the end of this narrow paddock leads to a further bridle gate opening onto an ancient and still well used rabbit warren in the corner of a small field, after which yet another unmarked gate for riders of horses takes one into more agricultural fields. The path here, although in a field, is obviously following the course of a very ancient track as witnessed by the bank on the left and banked hedge and ditch on the right; almost certainly an old boundary, and possibly still is. On the earliest Ordinance Survey maps this track is shown as a road of no less importance than that which it leads onto later. This probably accounts for the, now, double bend at Chapel Corner ahead.

More of the north east of the Island can be seen as you gently rise from these fields with the Ashey sea-mark and Brading Down most prominent. This is now cattle country, which is becoming rare on the Island, the gates become wider and you will wish for wellingtons! This is obviously a Wordsworth day: -

*The cattle are grazing,*

*Their heads never raising;*

*There are forty feeding like one.*

From 'Written in March'

The large farm coming into view on the right front is Mill Bank Farm. (So called because it's just above French Mill). Ignore any turnings down to the farm and continue straight ahead through two more large gates. The farm rarely uses the last section of track before the road, except as a dump, and a new bridle gate eventually takes you onto the pavement of the main Godshill to Shanklin road at Sandford Chapel corner.

Cross the road with great care (the traffic is very quick through here) and turn left on the other side. Note the sunken garden of the house now to your right, this was another excavation of clay for bricks, hence known as Brickfield. Now find the footpath (GL 33) to Bobberstone, on the right, between the thatched house and the tiled one a little way toward Godshill, before you reach the chapel. A nicely kept path leads up to a locked gate and unwired stile.

Over the stile and into the field; the direction of the path is not always obvious especially if the field is down to pasture. It is best to walk almost straight across with a very slight deviation to the left until you crest the rise, keeping well to the high side of the large concrete cattle trough in the centre of the field.

When the opposite hedgerow is visible it should be possible to spot the stile. There is a full half circle of views to the right from here; with, from left to right Arreton and Brading Downs, then Culver Cliff and the high land behind Sandown and Shanklin. This is followed by the ever rising ground leading to Shanklin and St. Martin's downs all the way round to the Worsley Obelisk and Gatt Cliff; with the Wroxall gap hardly discernible.

When walking across this field in mid-September it was alive with swallows and house martins skimming the ground to catch the last few meals before leaving us for their channel crossing. More recently no fewer than seven buzzards were soaring above; one wonders at the reason for the massive increase in their numbers.

The stile when it is reached is a steep double from this side. On the other side the path goes along the side of the field with the hedge on the right; this is an ancient hedgerow although, at the time of writing, it has been largely grubbed out and the saplings re-coppiced. **Approximately halfway between the stile just crossed and a stile leading through the hedge to the right, some two hundred and fifty feet further on, PLUTO crossed the path**. There is one of the Island's best preserved markers here. Always, as in all cases, providing it has not been demolished or disturbed before the reader finds it.

When the stile leading through the hedge on the right is reached do not turn right across the stile but make an acute turn to the left (About 150 degrees) and walk diagonally back across the field that you are in. This is footpath GL 31 but, at the time of writing, it is not signposted nor well-used. If there are crops the way will probably be marked; if not aim for the fields high spot and when that is reached aim for the far left hand corner of the field. Soon after you step off in this direction **PLUTO will, of course, have again crossed the path,** but as there is no hedge there would have been no need for any marker.

At the top of the field there are good views of part of the Island's central bowl. Especially prominent, to the extreme right as you get to the corner of the field, is Lessland farmhouse (a Domesday manor) standing tall and sturdy in its red brick amongst its arable fields.

In the corner of the field a stile leads into the horse paddocks of Froghill. From the stile there is a wonderful view of the Godshill area with the church, on its wooded knoll, dead ahead. Follow the field division on the right down to Froghill and at the end, by the gate, turn right and go down in front of the conifers. Beware of minor possible diversions around here and the use of electrified tapes used to control livestock grazing. The path used to go left over the little footbridge and cross in front of the farmhouse; but it now continues past the footbridge, still skirting the ditch, and past a further gate before turning left into a farm track.

This track continues for about seventy yards. Those interested in machinery might keep a sharp eye open around here as, amongst other things, there is often to be seen parts of a collection of Caterpillar tractors from D2's to D8's. These are not on display for the benefit of the general public and are not to be clambered upon or otherwise interfered with; but can sometimes be seen from the path. At the end of this bit of track you need to cross the main drive. Those interested could deviate to the Tack shop around to the left which also sells sport and country clothing.

After crossing the main drive, you enter a stable yard area but it is necessary to turn almost immediately right, aiming for the electricity transmission pole, at the base of which is a stile leading across a further paddocks keeping the fencing on the left. As you cross the field the Worsely monument on the top of Stenbury Down is prominent to the left. The stile ahead will take you, very steeply, down into Lessland Lane. You are now back in Sandford. Turn left here along the lane and proceed to the main Godshill to Shanklin road ahead. A right turn onto the verge will take you back, past the entrance to Redhill Lane on the left, to your starting point near Sandford House. Rather a devious route for two close together crossings but one has to make a reasonable walk of it!

## Walk 1

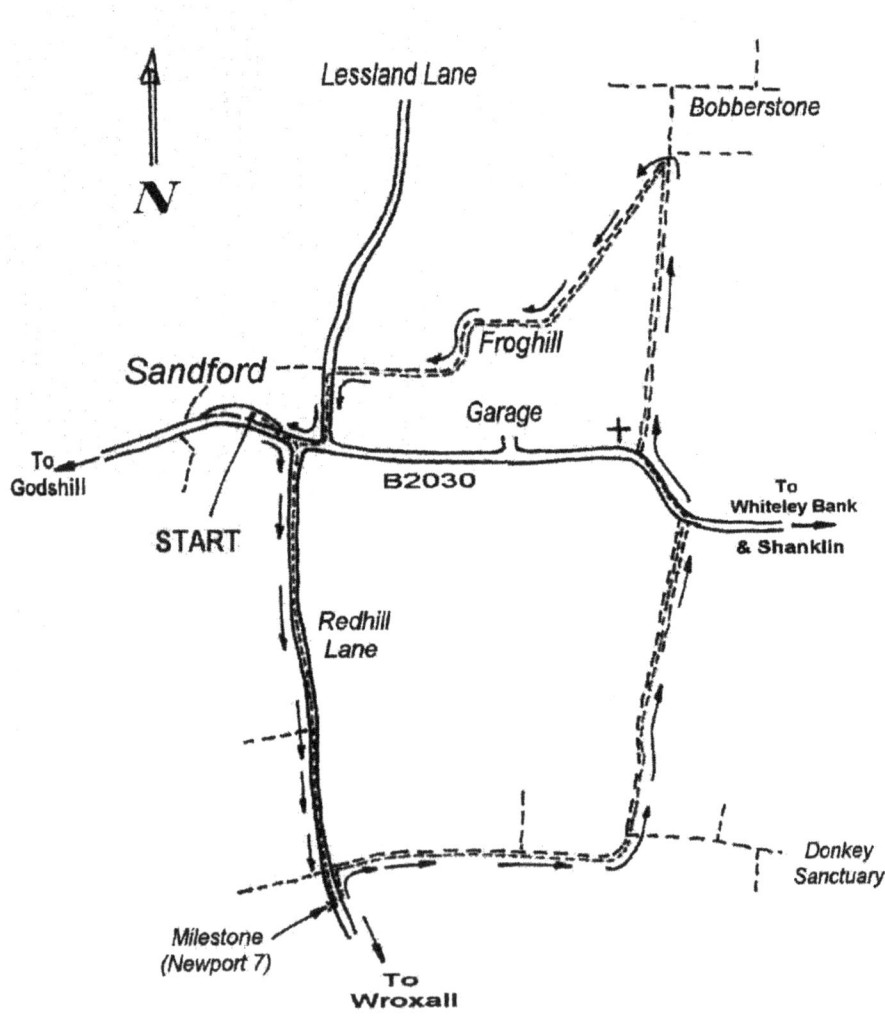

## 2. TO BLACKWATER AND ON TO ROOKLEY

*(A Wander South from Newport)*  *(About 6¾ miles.)*

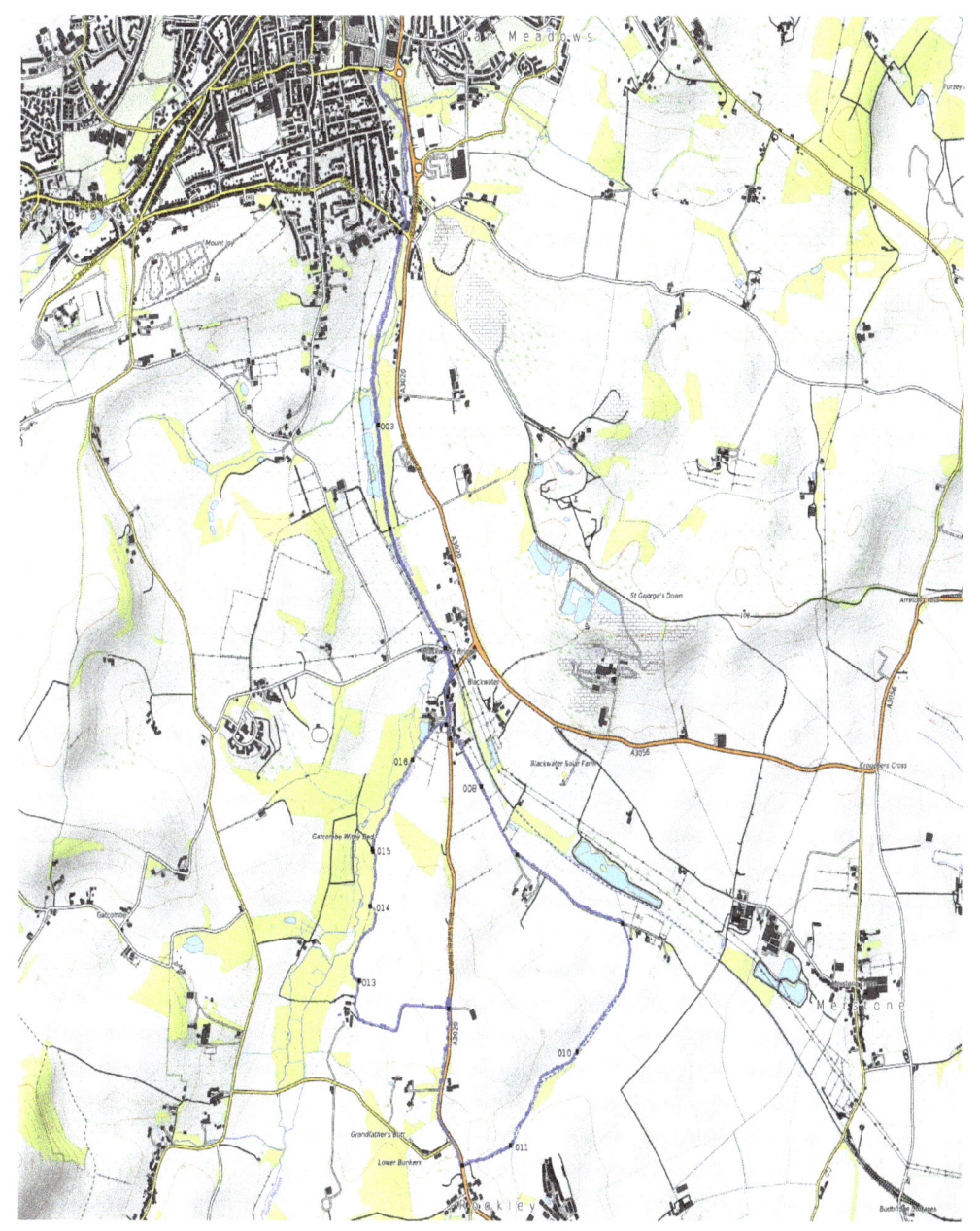

Leave Newport by St. Georges Approach but as soon as you cross the bridge over the river Medina there is a path, presently unmarked, by the 40 mph sign, which doubles back down to the river. Starting away from the bridge the path follows the river, the course of which was modified in the 1980s when St Georges Way was constructed. It was necessary to juggle the line of the new road with the river, the old railway, and the footpath. As you go over the small footbridge to the left by the first weir, the path takes you up onto the level of the old railway which the path now follows. Thirty years ago, apart from the copse on the left this was quite an open area; it's amazing how much the trees have grown in that time.

The people with gardens to the right have now made the most of not backing onto the railway, and to the left the road builders have incorporated banks screening this area from the road. Further on, when the area across the river to the right was being developed, evidence of the Roman occupation was found.

Also at about this point a branch railway used to go off to the left via a tunnel into Pan chalk pit; the cutting to the tunnel was filled in and made into a short road up to Burnt House Lane. That line also went over the millrace of Shide Mill which had to be demolished to make way for the new road. The chalk was used to make cement at the factory on the Cowes side of Newport. The millrace has been obliterated!

The poet Whittier has an apt verse:-

> *'No more a toiler at the wheel,*
>
> *It wanders at its will;*
>
> *Nor dam nor pond is left to tell*
>
> *Where once was Birchbrooke mill.*

From 'Birchbrook Mill' by John Greenleaf Whittier

At the end of the path turn right over the footbridge and then left, into Shide Path. The building, which now appears over the river to the left, is in almost exactly the same position as Shide Station was, and the opposite 'bank' formed the railway platform; the river, having been moved some forty yards west, now runs where the trains did. The canopy, nothing to do with the railway, was erected when it was a vegetable wholesaler's loading bay.

Cross the road and take to the cycle track, formally opened in 2003 but in use for many years before that. Once you get away from the new road bridge the cycle way makes use of the old railway track itself and what you are now walking on was, originally, The Isle of Wight Central Railway between Newport and Sandown; the new bridge is where the level crossing was. There is now a car park here; if you are travelling by car, it may be worth starting the walk here instead.

The older road bridge over the river was some distance east of its present position; about on the alignment of the new road running parallel with the river. Looking back, and across the road towards the chalk pit on the left, two circular pillars can be seen on either side of someone's entrance; these were part of the older bridge parapet, although not in that position.

Shide Bridge was, in medieval times, the meeting place for those organising the defence of the Island. Immediately across the river, between it and the road there was a row of houses called Malvern Terrace; they also became a victim of the new road, all that remains is one lone apple tree.

Slightly further on, new trees have been planted. Approximately in line with the last of the houses, over the river, is an Indian Bean tree which may be glimpsed between the willows. Almost alongside it, in 1974, the Japanese Ambassador planted a cherry tree in memory of Dr. John Milne the eminent earthquake scientist, who died in 1913. He lived at Shide Hill House which is behind and above those last houses.

To the right, in recent years, the land has been given over to the growing of Christmas trees but after that, at the time of writing, it looks rather 'set aside'! To the right is a white house known as Millstream Cottage, and then, rather screened by trees, the brick built Upper Shide Mill. The mill took water from the river further up and returned it to the mainstream after it had done its work; the millstream's exit, however, is no longer to be seen. (Almost another 'Birchbrook Mill').

A little further there is a small side path through the willows to a three-drop weir and, with a lot of luck, a possible glimpse of a kingfisher. Continue on the main track, the river is mainly unseen, meandering on your left. A few yards further, on the left, you come upon the remains of an information board and seat (now vandalised) which once overlooked a clearing and a, now unused, river gauging station where the water flows through a concrete channel of specified dimensions. It's all rather overgrown at the time of writing.

The little hut over the river that used to contain the equipment was converted into a bat roosting facility and no doubt now has other inhabitants. There used to be a footbridge here across to the hut which connected to a path across the field up to the Blackwater Road; not a great loss, it was always very marshy, but a lost path nevertheless.

The track now crosses a bridge where the watercourse from the right joins that on the left to make a single stream; more about this later. Through the trees to the right one can see the fields going up to the Blackwater road that, fifty years ago, were the home of post war County Agricultural Shows. A setting that I, personally, feel was more in keeping than its present one at Northwood; such is progress.

To the right, not visible behind high retaining banks, are newly formed fishing lakes and the view across to Marvel Farm is likewise obscured. Both sides are now are wonderfully natural habitats for all manner of plants and creatures well protected from two legged intruders by the river running either side of the track.

I think the rail engineers who built the line were very clever here. It was common to have generous ditches either side of the permanent way for drainage, but here they have two halves of a river. I believe that they quite deliberately split the river, as will be seen up ahead, and built their embankment up from the more solid bottom of the old riverbed rather than trying to support it on the alluvial deposits which occur on either side.

The path next crosses a track from the road through to Marvel Farm (not a footpath) with its small bridges either side. Soon after this, as you approach Blackwater, and after a small water extraction plant on the right, the end of this section of your path comes in sight.

There is a new bridge for the cycle path to the left and a footbridge to the right. The cycle way had to be diverted here, as the rail track and station (now a house) are privately owned. This is where the railway engineers split the River Medina, quite intentionally, to form their embankment, which, with the retaining walls, reduced the amount of scouring of banks and railway.

The route now goes over the small bridge to the right which takes you into Sandy Lane where you turn left and immediately go over a small, new, road bridge. Looking over this bridge to the right, the Medina can be seen coming in from the south and its junction with the 'Blackwater' (sometimes referred to as The Merstone Stream) coming in from the left with its own small depth gauge a few yards further on.

The roads in this area were rearranged when the railway was built. Sandy Lane used to go straight through what is now the converted station house so it had to be bent to the right to run parallel with the Blackwater Stream.

Sandy Lane takes you up to the main Newport to Rookley road; just up the road to the left, on the other side, can be seen an iron clad building that was Blackwater's C of E chapel of St. Barnabas, now used as a vegetable sales outlet!

Cross the road and turn right, the route crosses a bridge over the 'Blackwater' which gives the hamlet its more recent name. The date stone of the original bridge is still in position if you can spot it (1776, shades of revolution!). Before then it was a ford, hence the previous name for Blackwater which was Huffingford (or Huncheford in the Domesday Book); why it was changed and when, I've yet to discover except that it has apparently been Blackwater since at least 1548.

Quite an industrious little place, Blackwater once had a lace making activity, probably an offshoot of the Newport industry. There was a tramway from the top of St. George's Down with which to bring down the gravel. It has had a building firm since 1790.

As you leave the bridge you pass the site of a garage (currently being rebuilt), a chapel, a blacksmith's forge and the old post office; none now used for its original purpose. The forge and Post Office both joined to an architecturally unusual cottage. On the opposite side of the road as you pass the newer houses is the entrance to Blackwater Mill, the second mill on the Medina; now a residential home. Note the round pillars either side of the entrance; more parts of the old Shide Bridge parapet!

Continue along the road, and look for a track to the left; not the footpath (A39) but the bridleway (A36) about twenty-five yards further on; (The 'A' prefix to these paths denotes Arreton Parish which you will have been in since soon after leaving Shide)

This bridleway (part of the Stenbury Trail) is one of the old lanes leading to Merstone Manor (there aren't any new ones). The lane is fairly uneventful in itself, the cycle track, having regained the line of the old railway, is parallel with the lane across the field to the left, and behind that, St. Georges Down is now visible. To the right can be heard, rather than seen, the traffic tearing up and down Blackwater Shute. Apart from the traffic noise which is soon left behind it is now just an enjoyable walk through the countryside.

After about a quarter of a mile the cycle way joins the lane from a track on the left having had to leave the line of the old railway due to its non-recovery from the landowner. (Stopping the trains was bad enough, but what a huge mistake to have sold off the permanent way, into private ownership!). You now have to, again, be wary of cycles coming from behind with no bell!

The track now forks to the left, leaving Birchmore Farm to the right. The farmhouse, up to the left, has long since had nothing to do with the farm buildings a little further on which, sadly, are in the process of being converted to dwellings.

As the lane becomes clear of the trees into more arable land St Georges Down is again very prominent to the left and if the wind is right you can hear the gravel washing machinery which is clearly visible on the skyline. A line of small lakes, constructed in recent years, breaks up the farmland in between. The area between these lakes and St Georges Down is known as Long Down. The two properties visible are on another track to Merstone Manor which runs down diagonally from the main road.

Merstone Manor itself, built in red brick around 1615, can now be seen to the left front across the fields; easier to see in winter than in summer! Walking here recently I noticed that the hedgerow to the right has again been beaten to death, rather than cut, by some flailing machine quite unsuited to the size of much of the growth; sadly, now an all too familiar sight.

The next turning point comes up to the right just before the trees and the next property, Little Birchmore, on the same side; this is a bridleway to Rookley (A37). NB. You leave the Stenbury Trail here. You are in more open country now with a small stream and hedge to the left; strangely the hedge gets older as you progress, as witnessed by the increasing number of species making up the hedge! The initial lengths must have been replanted at some time.

The path rises steadily and after about half a mile you come to an interesting little copse preceded by a wonderful old oak tree, there are others to follow but none quite so venerable as this.

I say an interesting copse because, to me, it has a 'presence' of times past, I cannot believe that at some period in the past it hasn't been inhabited; it has water, wood, grazing and fruitful soil around about and is in a nicely sheltered position.

The path runs alongside the copse and although narrow, the copse is quite long. Peering into the copse it will be seen that in places, due to the rising ground, the stream has cut a relatively deep little ravine. Whilst undoubtedly fed partially by field drains and ditches this little stream appears to start at a spring, perhaps somewhere near the old Rookley brickworks.

At the field boundary the definitive path actually goes diagonally across the field to the opposite corner. This seems quite pointless and I am sure everyone walks around the edge. So, after the field boundary, the path continues to skirt the copse and starts to curve to the right, passing a broken bridge and gate at the end of the copse on the left (this isn't a footpath).

**Very shortly after leaving the Stream, little more than a ditch now, and turning more definitely to the right PLUTO crossed the path, its contents flowing from right to left!** remains of a marker can be found here, but only with some difficulty, as there is much Ivy.

The walk now takes you up, again, to the main Newport to Rookley road where you emerge opposite another iron clad 'hut' that was originally Rookley's C of E chapel of St. John. This had until recently been used as the village hall but Rookley now has a nice new one (visible shortly through the trees to the left).

Cross the road and turn right; unfortunately, you have to use the road for a while. Initially there is a reasonable path along the roadside but after passing Highwood Lane, on the left, it is difficult to avoid walking on the road. Personally I would stay on the left-hand side, there is at least a refuge to be had on the verge in places, while the other side is completely blind to traffic coming from your front. (Take especial care on the bend).

On the left after the bend is Pidford Manor; an unusual house or rather two connected houses. The front, facing the road, was built in 1795 but that at the rear is older and believed to have been built around 1630. Both built by the Worsley family who owned the estate for nearly three hundred years until 1825.

After passing the entrance to the Manor and the farm opposite and keeping to the left, **you come to a point where PLUTO crossed the road**. Until 2010 there was a footpath here, the A41, going up the bank to the left and across the field, but the landowner objected and achieved a diversion. However, if the foliage is not too dense you may spot the single post of a marker at the edge of the field at the top of the bank, **but note, PLUTO crossed the road here,** not the footpath. The road here, and its cutting, has been considerably widened since 1944, so the marker which would have been on the east side of the road has long gone.

To continue the walk, you will now have to go to the top of the hill where you will find the FP sign on the left pointing down the drive to Champion Farm. There is a fine view from here, to the left above Pidford is Bunker's Hill, then a glimpse of the hill above Loverstone, followed by Chillerton Down with its TV mast which rolls in succession into Newbarn directly ahead. Then there is Garston's Down, over the shoulder of which peeps the Rowridge TV mast. In the middle ground at that point stands Gatcombe House, one of the Worsley built mansions. Following the view to the right, somewhat further off is Alvington Down then the dip in the Carisbrooke direction with the water tower of Whitecroft asylum in front being overlooked on the extreme right by Great Down.

Continuing then, down the drive which, after a while, takes a left turn and joins up with the line of the old FP upon reaching the hedgerow when it turns right and continues along past the Champion Farm boundary continuing to bear right. There is a bridleway/footpath to the left just before this and after the farm, not to be taken! When that point is reached you continue with the hedge on your left. Continue to follow the boundary on the left which now separates you from the wooded marshland of the river Medina. It is not signed but you are now on path A40 which for no obvious reason becomes A42 before we get to its end. The path carries on around the edge of the field. You will come to a kink in the path to the left, it used to go straight on here to a cottage called 'Gilmans'. However, you need to carry on following the field boundary.

**Somewhere between this kink in the path and the next field boundary PLUTO crossed the path, this time flowing from left to right. I cannot find even a hint of a marker; can you?**

I am intrigued as to how they took PLUTO across the river valley here; was there enough cover in 1944 to run it above ground? Where it crossed smaller streams in more open country it was taken beneath the water.

After that field boundary coming in on the right, the path continues for a short distance to hug the edge of the woodland before actually diving into its edge for a while. The site of the now, non-existent, Gilman's Cottage was a few yards into the field on the left (this field was a wood until a few years ago). It is possible to skirt this wood, but appears to go through a shooting range – best not.

I suspect that at some time Gilman's was a gamekeeper's dwelling because after leaving the little wooded section of the path and reaching the top of the next rise, there are some ruins (in Ivy) to the left, also known as Gilman's, which include the remains of some purposeful looking dog kennels.

As you reach the last of these very overgrown ruins (they are quite small and not easy to see) the path joggles to the left again before continuing downhill; ignore the track into the woods to the left that leads into Gatcombe Withy Bed which is private shooting country (OK I suppose unless you're a private!). Walking down here in the autumn is a feast for the eyes there is such a variety of trees and saplings, the colours are wonderful. It's also pretty good in summer.

At the next field boundary, a stile at the time of writing, the path becomes fenced on both sides. This has happened sometime in the last fifteen years. There has been some sympathetic landscaping and tree planting to the right of the path; traditional farming in the area having dropped off. Now, for a short distance, the river Medina runs alongside the fence to the left; there is a rustic old bridge that appears to go nowhere, and is well hidden in foliage even in winter, and soon after comes the start of the 'management' of the river for its entry into the mill pond of Blackwater Mill.; now a residential home for the elderly.

The rusty ironwork top of some old hatches can still be seen a little over to the left. These were vertically sliding 'gates' which could be raised or lowered to control the flow of water into the millpond or force it to flow into the by-pass channel; especially in times of flood. After that it is all quite new.

The millpond has been converted into a very attractive ornamental lake for the benefit of the residents, and no doubt the owners. This lake is visible through the high conifer hedge as one continues down the path which now twists around the stables at the end of the field and emerges onto the Newport to Rookley road alongside the entrance to the mill. There are those bits of the old Shide Bridge parapet again. It is probably sensible to cross to the pavement side of the road before turning left and retracing your steps back to Blackwater Bridge. At the bridge recross the road again into Sandy Lane, on the left. After crossing the small bridge over the Medina, in the lane, there is a choice of two routes back to Newport. You can turn right into the cycle track (the old railway) and retrace your steps back to Shide.

Or, you could continue up Sandy Lane and take the first turning right which takes you into Newport via Marvel Lane and Watergate. Ignore the turn off to the left, just after the little bridge at Water gate and keep to the main lane, which bears right. The lane terminates at the crossroads known as Shide Cross (presently a mini roundabout) when St. John's Road, opposite, will take you to the town centre.

**Walk 2**

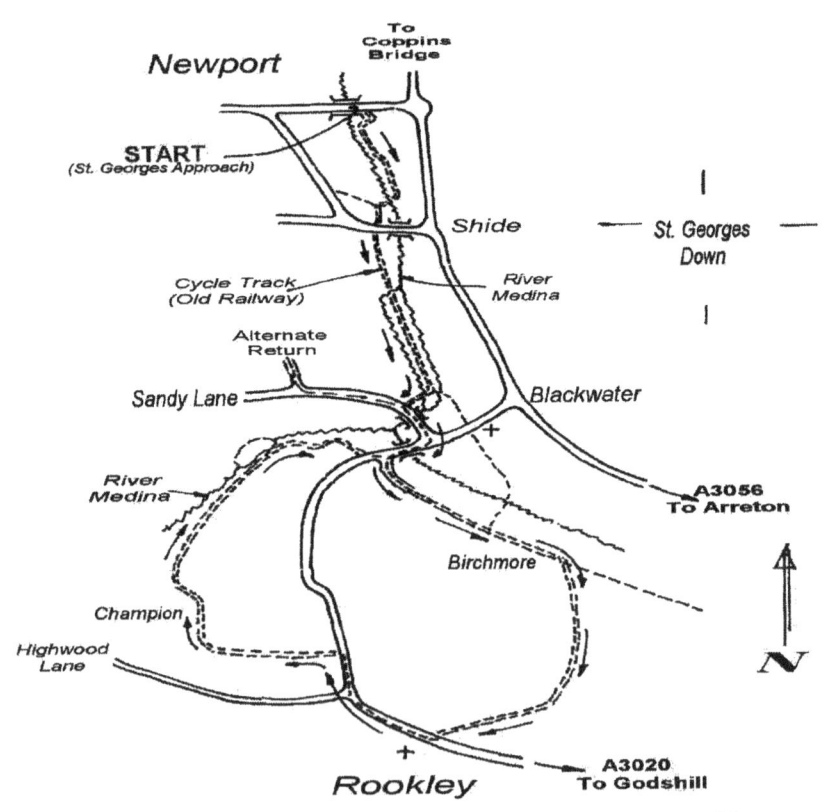

*Not to scale*

# 3. A SHORT SHANKLIN CIRCULAR

*(Not a walk, just a casual, after lunch, stroll)*     *(Just over 1 mile.)*

This short walk will take you across the PLUTO pipeline that ran from the Hungerberry wood storage tank codenamed TOTO, down to the Shanklin pumping stations.

This little walk takes in an interesting little bit of old Shanklin. I won't attempt to delve into the origins of the Shanklin name, there are too many theories, none very convincing; suffice to say that in the Domesday Book it's given as Sencliz or is it Selins (the experts can't agree).

Shanklin started to become popular in the early years of the nineteenth century; Keats was one of the first to write of its undoubted beauties, he talked of its greenness and of 'the sea Jack the sea'. Other persons who became well known followed; as you walk you may be treading in the footsteps of H. M. Stanley, who came here for a rest in 1894; long after he had tracked down Livingstone. It was also well known as a place of retreat for university reading parties.

From the old village take the road going inland opposite the car park, this is Grange Road, almost certainly so named because 'from ancient times' the tithes of the manor were payable to Quarr Abby. At the top of Grange road turn left into Pomona Road, leading shortly to the start of West Hill Road and, on the left, is Rectory Road where once was Hoares Brewery and later a wine bar, now, regrettably, a housing complex.

Almost immediately after Rectory Road, on the same side, a lane goes off at a tangent, this is Manor Road although it doesn't say so; it is only signed as a footpath (SS84) to 'The Manor House' and to 'The Old Church'. It was originally the carriage drive to the manor (and, before 1829, it was also the road to Bonchurch). It later became a gated road.

The path, now metalled, is between high banks with woodland on the right and, through a narrow screen of trees, a large open field on the left; it's a nice, relatively unspoilt area. The Town Council had the foresight to purchase the whole area in 1933 and opened it for public use (unlike recent Councils who are continually selling 'the family silver'! A 'one off' activity). A recent eyesore in this large open space is the garish fence surrounding the children's playground; the need is understood but surely more sympathetic colours could have been used.

**To return to the path; as you pass the playground, PLUTO crossed the path**. There are some remains of markers either side of the track, one side more obvious than the other. Two years ago, in 2002 the crossing point could be seen in the surface of the path but it has since had new tarmac.

Continuing up the path it is noticeable, even at the height of the holiday season, how few people use this area; it is really quite special, a calm, so unexpected this close to a major grockle centre. There are some superb trees here, many of the tall straight ones probably dating back to at least 1880, some no doubt earlier. With tall trees I always think of part of Thomas Hood's poem 'I remember I remember'.

> 'I remember I remember,
> The fir trees dark and high;
> used to think their slender tops
> Were close against the sky:
> It was a childish ignorance,
> But now 'tis little joy
> To know I'm farther off from heav'n
> Than when I was a boy'

It's now difficult to imagine that the open space on the left known as 'Big Meade' was a mere field as recently as 1932 when, we are told, cattle were still being 'fattened up' on it.

The walled garden of the manor soon appears on the right followed by the manor itself, now mainly a Victorian version but, nevertheless, quite handsome. Some tall, rather unusual, outbuildings nearer the path are earlier and are listed due to their being used for important scientific experiments.

The title of 'manor' dates from the holding here at Domesday. In 1845 it was referred to as a farm and pictured as a large square house although it must have been quite an establishment in the mid 1700s when it was known as a centre of resistance to the Hanovarians and toasts were drunk to 'Charlie over the water'. Not surprising, one hundred years earlier it had been in Royalist's hands. The present building was built around the core of an earlier, possibly 17c house, for the White-Pophams during 1883/9. The Council bought it in 1933 but sold to the WTA in 1935. It was a hotel but has now been converted into luxury apartments. Some of its accompanying outbuildings have been converted to dwellings and some, in the area of the farm, predating 1883, remain (in 2004) as ruins but these are now being rebuilt.

As the entrance to the hotel is reached your path turns left along the hotel entrance drive, still known as Manor Road. To the right some of the ruins are being incorporated in a new walled garden. To the left a much older pond appears with a variety of waterfowl; this is the western termination of Big Meade.

Your route turns right at the end of the new wall but it is worth a very short detour here to have a look at Shanklin's oldest church. The dedication was to John the Baptist but later changed to St. Blasius (which had been used for a while in the 1300s). It was originally built as a chapel to the manor by the De Lisles (or De Insulas), first as part of Brading parish and later, until 1859, as part of Bonchurch. It has, of course, been much modified by the Victorians who have often been criticised for what they did, but it has to be remembered that many churches were near ruins until the prosperous Victorians got to work.

Before you leave the churchyard go to the extreme left hand side of it and, with care, look how deep the chine is at this point; the fact that the chine extends as far inland as this is often overlooked. There is a much better view from the road outside!

Getting back to your route. From the church regain the path, to the left now, at the end of the wall which is marked SS85. The entrance to this path is presently somewhat more clinical than of late but it will age and be all the better for the clean-up. Follow the manor wall round past the converted stables and you arrive at a wonderful open space with many mature trees; the path becomes an avenue between great ash trees.

Ahead and to the right, across the grass, is the entrance to Shanklin's cricket ground; a peaceful setting and worth a pause, even if they're not playing one can imagine the sound of the proverbial leather upon willow.

Before WW 2, and for a few years afterwards, this ground consisted of lawn tennis courts where matches of a high standard were played. It did not become Shanklin's cricket ground until 1951.

**Continue down the path; almost immediately after passing the cricket ground entrance we enter an avenue of laurel bushes and, in a very few yards PLUTO crosses the path.** There is a marker, well back from the left side of the path, although it is no surprise that one on the cricket ground side has not been found.

If you look across the fields to your left from here, you will see Hungerberry Copse whose trees and bushes along with much camouflage netting concealed TOTO, the 620,000 gallon header tank from which the fuel was gravity fed through the pipe to the chine and on to the pumps.

The path ends upon entering West Hill Road, turning right allows you a pleasant walk back to the Old Village passing, on the left, some nice properties, some old, some new. On the right is the cricket ground and then other open spaces. You soon reach the entrance to Manor Road and can retrace your steps back to the Old Village through Grange Road noting the rather grand 'Old Church Parish Room' on the right.

**Walk 3**

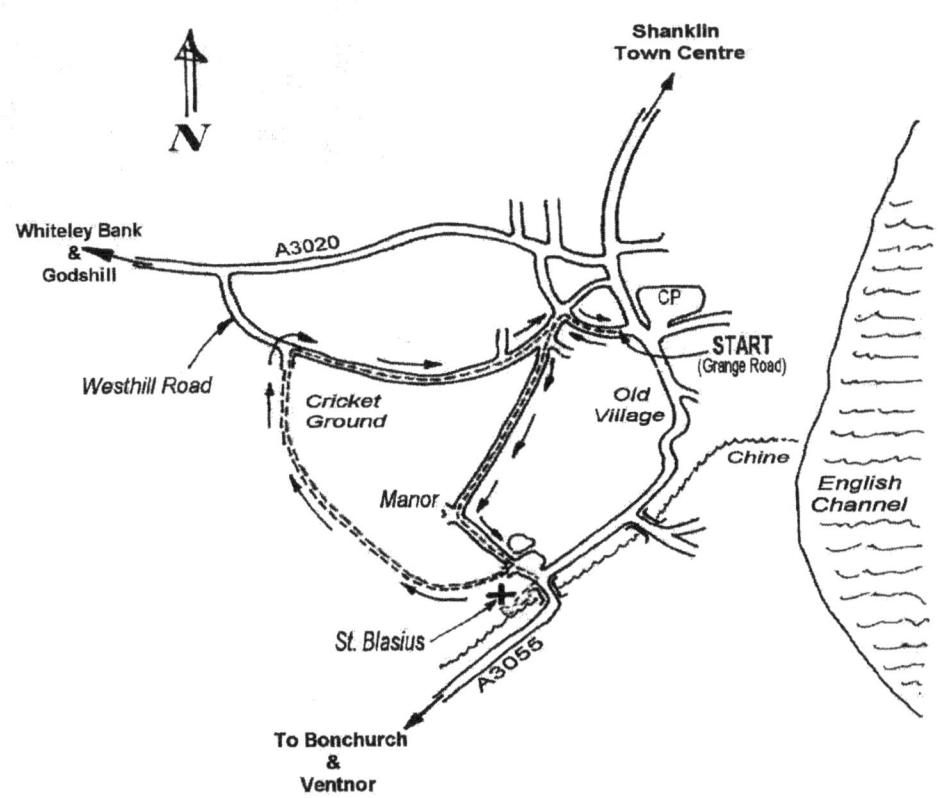

*Not to scale*

# 4. MARKS CORNER TO THE SOLENT SEA

*(A corner of Parkhurst Forest, plus some) (About 4.5 miles)*

Marks Corner, in terms of getting there, is to the SW of the road between Noke Common and Gurnard. If coming from Newport, turn left at Chapel Corner, Parkhurst, (this is Noke Common) and follow the road that now forms the north-eastern boundary of the forest. Turn left off that road, after about one and one third miles, at Hillis Gate.

The hamlet is not of manorial origin but is by no means modern and has a little history that is all its own. It is untouched by development in the recent sense although there are a few new houses, some possibly replacing old ones, and most have been attractively modernised. The one time 'Primitive Methodist' chapel has been converted into a dwelling but to me Marks Corner remains a jewel in an otherwise overdeveloped Island. It is very easy to be envious of the peace of these natural surroundings.

With reference to that chapel; somewhere along the way the WW1 memorial has disappeared. This is always a worry when churches and chapels are de-commissioned. If anyone is aware of its whereabouts please contact someone at Marks Corner; Bradley, Flux, and Thompson are just three of the names on it. (None of them must be forgotten).

Notice that it is not, or need not be, Mark's Corner. Like many of these names the origin is obscure and no one is sure if someone called Mark was involved or if indeed the place was a mark! There is a small car park and a bench, right on the corner.

So, to walking. Starting westerly from Marks Corner with the houses on your right and the forest on the left, you will, near the end of the houses, come to an entrance to the forest and a footpath signpost for CB5 to 'Whitehouse Road' pointing to the track ahead. After about forty yards the track veers into properties to the right, including the 'new' entrance to Stagwell Farm, but the footpath goes off slightly to the left, still heading west. Note the CB; for all of this walk you will be in the parish of Calbourne.

The path runs along the border of the forest with boundary banks and ditches on both sides which are a long way apart. So far apart that you may not have noticed them as the path is now quite narrow. These wide boundaries denote an ancient way which, as late as 1901, the council proposed turning into a modern road to Porchfield. Fortunately, it didn't happen. The bank on the right has many fine oak trees defining that boundary, probably dating from the early 19c when the Ward estate was being expanded, although the boundary itself must predate that, as Stagwell farmland was cut out from the forest many years before the Wards came to the Island; possibly in the 16$^{th}$ century or even earlier.

Due to the centuries of use without modern foundations and the use of various bits and pieces of hard-core this can be a rough path in any weather; sometimes more of a stream. After about three hundred yards you will come to another entrance to the forest on the left, this is known as Stagwell Gate; the wooden arch is recent. On the opposite side of your path is another footpath going off to the right; this is said to be the old entrance to Stagwell Farm, and there is evidence in that field of more of a track than a path.

However, your route lies straight ahead and after a short distance the path, with its real boundaries still wide on each side, begins to turn more into the forest to the left, slightly downhill. This is to skirt an area to the right at the bottom of the slope known by the strange name of Dogs Ant. There were some cottages here but they fell into disuse, before WW II apparently as the result of a fire, and were eventually torn down. Beware if you are tempted to poke around here, at least one of the wells remains; heavily camouflaged by undergrowth but not covered! The name has to do with a feeding area for deer and hunting dogs.

For those who may feel uneasy when walking in woods alone it's worth remembering a couple of lines by William Barnes: -

> 'In whatever season it mid be
>
> The trees be always company'

(The Dorset dialect is not far removed from that of the Island)

The path continues. As will be seen by the remains of old stiles by the wayside, the path, whilst still between the overall boundaries, has had its course varied over the years as people have taken the easiest route when it became overgrown. Eventually, the path, after crossing a small brook in the margin of the forest, brings you, by way of a stile, into open (& muddy) fields. Your route goes across the field ('Nine Acres') keeping to the left-hand hedgerow. (My field names are taken from a document originating in 1815. It is possible that some may have changed. Some field boundaries certainly have!)

**At some point between that stile and the pole for the overhead power line ahead PLUTO crossed the path.** I can find no marker. Unfortunately, other pipelines, including the most recent one for gas, have used this route since the war to bypass the forest, just as PLUTO did, and have almost certainly destroyed the marker.

A stile takes you into the next field ('Six Acres') and the other side of that is what is said to be Little Whitehouse Road. (This is the Cowes to Shalfleet road), Whitehouse Road, proper, starts about twenty yards to the left where the main road takes a sharp right to Porchfield. It is interesting to note that the alignment of the present road to Porchfield suggests that the old track which your path followed through the forest would have emerged in line with that road, when these fields were still part of the forest. Probably long before the 1793 survey and reduction of the forest in 1815. At that date the main track from the forest to Porchfield was on the line of the present footpath from Whitehouse Farm to Porchfield; a little to the south. So it appears to have reverted to, perhaps, a much earlier alignment over Bunts Hill!

In order to pick up the next footpath, which happens to be the Coastal Path, it is necessary to use the road. It would be possible to take the road to Porchfield and take the first footpath on the right but I consider that bit of road too hazardous. So it's over the stile and turn right onto the road; at least in this direction there are some verges to take refuge upon.

The houses on this part of the road are considered to be part of Porchfield although that village is still nearly a mile away to the south west. This road is the 'back road' to Cowes and as such is busy and the traffic surprisingly quick, so beware. As you leave the houses behind there is a view, to the left, of South Thorness Farm and the beginnings of the Thorness Bay Holiday Camp, or as it is now known, Holiday Park, which is your next destination. Over the rise the Solent comes into view with Fawley oil refinery on the mainland ahead. At the bottom of the slope, as the road takes a sharp right, turn left into the lane to the Holiday Park. This junction is known as Thorness Cross. There is a seat here and the remains of a stop for the, now defunct, once a day bus. The 1862 map refers to the buildings on the left as a school.

The lane goes down to a small bridge over a stream, you can admire the well-built, brick culvert if you peer over the side. (only of course if it is of any interest!) After this, Great Thorness Farm comes up on the left. Near the top of the hill go past the entrance to the Park on the right. Shortly after this the coastal path (CB12a) joins the lane from the left and opposite this is the centre block of the original holiday camp with its rather 'art deco' tower. Continue up the lane which now starts to bear round to the right; on the left is South Thorness Farm followed by the farm houses. Not a lot of farming is done here now, only horses and caravan holidays. After the campsite on the right the next part of the Coastal Path CB24 takes off from the lane on the same side. The sign is half hidden by some fir trees. In muddy weather it is perhaps better to have taken the road through the Holiday Park, as this path can be difficult.

The name Thorness is the result of the usual confusion. From early days it is said to have been to do with a thorny hedge, (Thorneye in 1285), but somewhere between 1400 and 1750 the 'ness' bit appeared; referring to the headland!

The path now goes through a meadow, somewhat 'set aside', bordering the campsite and then by means of a stile into a wooded part of the caravan park. Turn sharp right after the stile to avoid going between the caravans. A short path, with some stumps to the left separating it from the caravans, soon joins the main tarmac-surfaced roadway through the park. Turn left onto this – or continue, if you used the tarmac road to get here. Head downhill between the chip shop and the Entertainment Centre after which will be seen a gravel track continuing to the beach.

Thorness Bay is one of the few places between Gurnard and Newtown on the Island's north west coast where access to the beach is practical. It is possible to walk the shoreline between Gurnard and Burnt Wood providing the tide is right but beware, the treacherous 'Blue Slipper' clay is not to be dismissed lightly. It is extremely slippery, can be very cloying and will stain any clothing beyond removal. (Burnt Wood is accessed via a lane and footpath from Porchfield)

Continuing on the gravel track to the beach a good view is to be had off the high ground above Gurnard, to the north west; the lower slopes dotted with the occasional private huts, some dating from the 1920s. The chimney prominent on the mainland, to the north, is not part of the refinery but is for Fawley power station and is, surprisingly, 650ft high. That's nearly 250ft higher than Salisbury Cathedral and almost 90ft higher than Portsmouth's Spinnaker Tower!

The gravel track down to the beach passes through meadows on either side that have been left to return to nature. On the beach continue to the right, that is north easterly, past the posts of the vehicle barrier and keep generally to the high part of the shingle. There is a large reed bed habitat inland to the right. This is one of only two occasions on this series of walks where you could indulge in a little beachcombing.

It was not always so peaceful here. On this beach and on those low cliffs to the north side of the bay, in the early summer of 1944, troops of the First Guards Division were practising for the D-Day landings. The bay itself was crowded with landing craft and ships, either taking part in the mock landings or straining at their anchor chains, waiting.

Ignore the stile on the high bank to the right, that is for footpath (CB3), which leads up to the road via Little Thorness farm, and is not a part of this route. As you come to an area of salt marsh don't be seduced, by the map or the appearance of a path going away to the right by the hedge; it only goes to an old sluice. There were salt pans here in the 1700s. Keep to the left (but not too far!)

There are some small boats kept and launched from here but it is not a facility available to the general public; those who use it will have made some arrangement with local landowners. The shingle spit is an excellent place to look for waders and other bird life.

Eventually, it's not very far, you will come to a small concrete bridge over a minor inlet (part of the old salterns) which leads to a track going inland. The Coastal Path carries on atop the low cliff but you leave it at this point. **It is well worth taking a short detour along the beach a little further; especially at low tide, there is a large group of old pipes (codenamed SOLO) coming ashore to a series of wooden posts in the sand, this is the manifold where 20 fuel lines converged on the beach to make the single PLUTO line crossing the Island.** Erosion is steadily taking its toll of these. Return now to the concrete bridge.

The path inland is signposted as the CB2 to Whippance Farm and Rolls Hill. After a short distance a track goes off to the left, ignore this it is another way onto the Coastal Path via some of those private retreats mentioned earlier, some old some new. **It is at about this point that PLUTO crossed the path after coming across that cliff top meadow to the left (see walk No. 17) to take a, seemingly inexplicable, wide sweep around the far boundary of the field coming up on the right.** This was undoubtedly due to the pumping station that was located in the now partly flooded area to your left. The curve of the pipe allowed the high pressure to stabilize as it started its 14-mile journey across the Island. No markers are to be found; it is possible that there never were any this close to the pumping terminal!

**Continuing inland, on this well consolidated track you are quickly into centuries old farmland, but before you reach the end of the next field to the left PLUTO again crosses the path.** This time from right to left across that rough ground bordering the track; having swept up the curving hedgerow, and ditch, of that field to the right. I have found no markers here but the hedges are dense, so there may be some!

Whippance Farm now comes up to your right at an apparent crossroads but there is no footpath left or right. The house does not seem particularly old but the location is of some antiquity. It may once have been, indeed may still be, connected to Sticlett. It is another odd name said to be perhaps a family name but I do wonder if it has anything to do with the bar which was used to yoke two or more oxen to a plough; that was called a Whippance!

After the farm the track dips to cross a small watercourse before rising again. The hedges are noticeably old as witnessed by the variety of the shrubs in them. Note also the thoughtfully angled gateways into the fields on either side, making it much easier to get vehicles in and out. This was especially important when it was a Hay Wain drawn by two or more horses; not an impressive turning circle on those!

From the high point on the track the green fields ahead can be seen sloping up to the forest and prominent, slightly to the right of centre, is Stagwell Farm, another old and somewhat remote farmstead. **From before Whippance Farm and as you approach the road ahead, you will have been walking parallel with and quite close to the line of PLUTO, which ran along the other side of the hedgerow to your left.**

**At the road, which was the one you walked along further west, turn left and, as you do so, PLUTO would have immediately crossed your path.** There are good markers here on both sides of the road – those on the north side appear to have been moved, rotated incorrectly and placed too close together.

Please be very careful here, there are no verges and the traffic is fast and furious. Fortunately, there is only about one hundred yards of road to suffer before a footpath takes off to the right with a stile and a sign saying CB4 to Marks Corner. This path, eventually, goes slowly upwards through three fields, with two more stiles, alongside Chalkclose Copse on your left.

The fields are called Lower and Middle Redlands; the third field and the one beyond the copse both seem to be named Bran, or Bean, Butt.

At the end of the third field a stile, in the top left hand corner, leads you through the neck of the copse into yet another field. Ignore what appears to be a stile in the far right corner and keep to the left around a treed, but quite open, area which terminates near the garden of a seemingly rather remote house. A stile here puts you onto the track to Stagwell Farm, which is away to the right.

There is a path through Stagwell Farm, thankfully not yet diverted. Although not on my route, you could, as we have finished with PLUTO, return that way to Marks Corner as an option. Stagwell is yet another place with a name to play with. It is said not to be as obvious as it seems. It's admitted that the 'well' is water in some way but is suggested that the first part may be to do with a marker post (stac). It depends how it started out, as now spelt it's tempting to think of it as where the deer of the forest came to drink!

(If you do opt to go that way; go through the farm, and paddock beyond it, then keep left up and around the pond with the hedge to the left and you will join your original track through the edge of the forest. Go left from there, retracing your steps back to Marks Corner).

However, back to my route, turn left onto the track, past the front of the house, and about eighty yards on, opposite the house, is a stile into the field on the right. There is no defined path, it seems little used, just continue uphill close to the hedge on the left. This is Lower Heath.

In times long gone these fields must have been cut out of the woods or heath with much hard manual labour, it makes one wonder how long they would take to revert back if left as lazy 'set aside'. In the top left-hand corner of the field is a stile. Looking back, what a view; the Hampshire coast from about Sowley in the west to the mouth of Beaulieu River in the east, the Solent itself and, on the Island, the slopes of Thorness Bay valley with its pastures sweeping down on either side.

Turning your back on the view, which may be difficult, cross the stile, go alongside the pond behind it, leading into a meadow (Upper Heath). Again keeping the hedge on the left continue past the yard of Jubilee Dairy and exit through the farm gate to the left. The footpath, by means of a stile, now actually goes through the duck and geese pond area to the right; but only for some ten yards. This is all rather pointless as it is only a few steps to the road where you would end up after going through the pond area! This calls for a diversion that would be sensible. Turn right up the road and you are back at Marks Corner with its interesting collection of dwellings.

## Walk 4

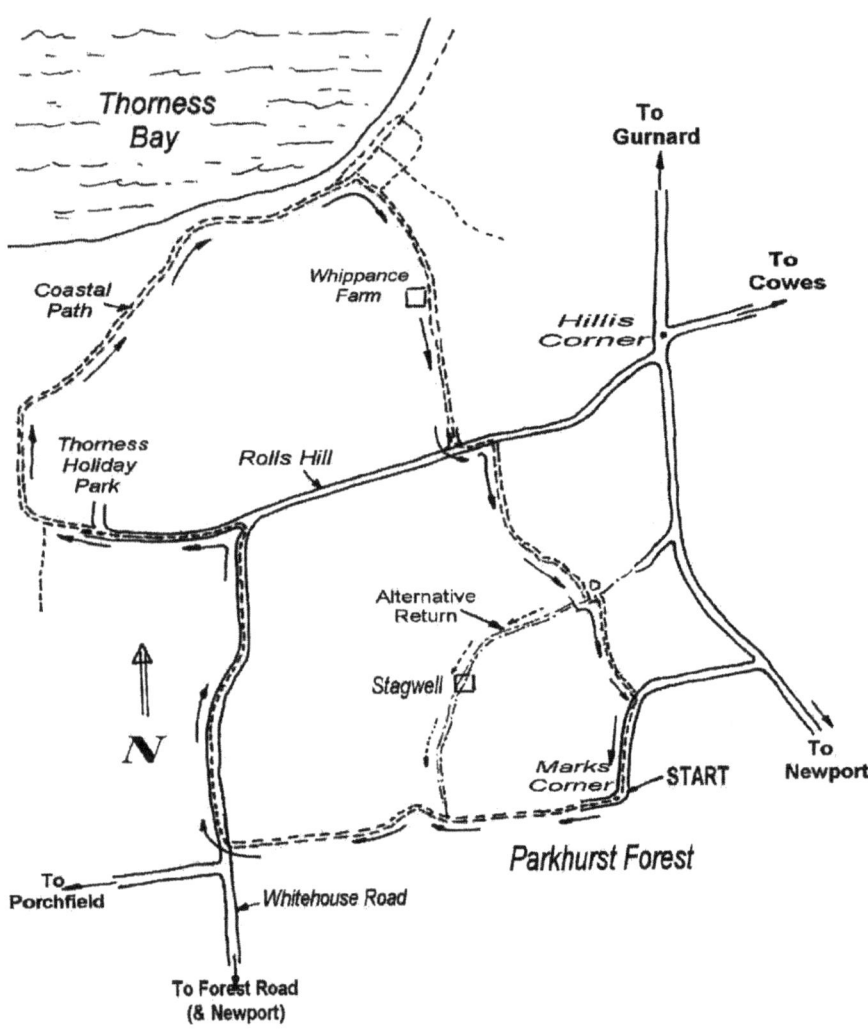

*Not to scale*

# 5. BOWCOMBE VALLEY AND THE LUKELY

*(A walk south west of Carisbrooke)*

*(About 2¾ miles. or with extension. 3¾\*)*

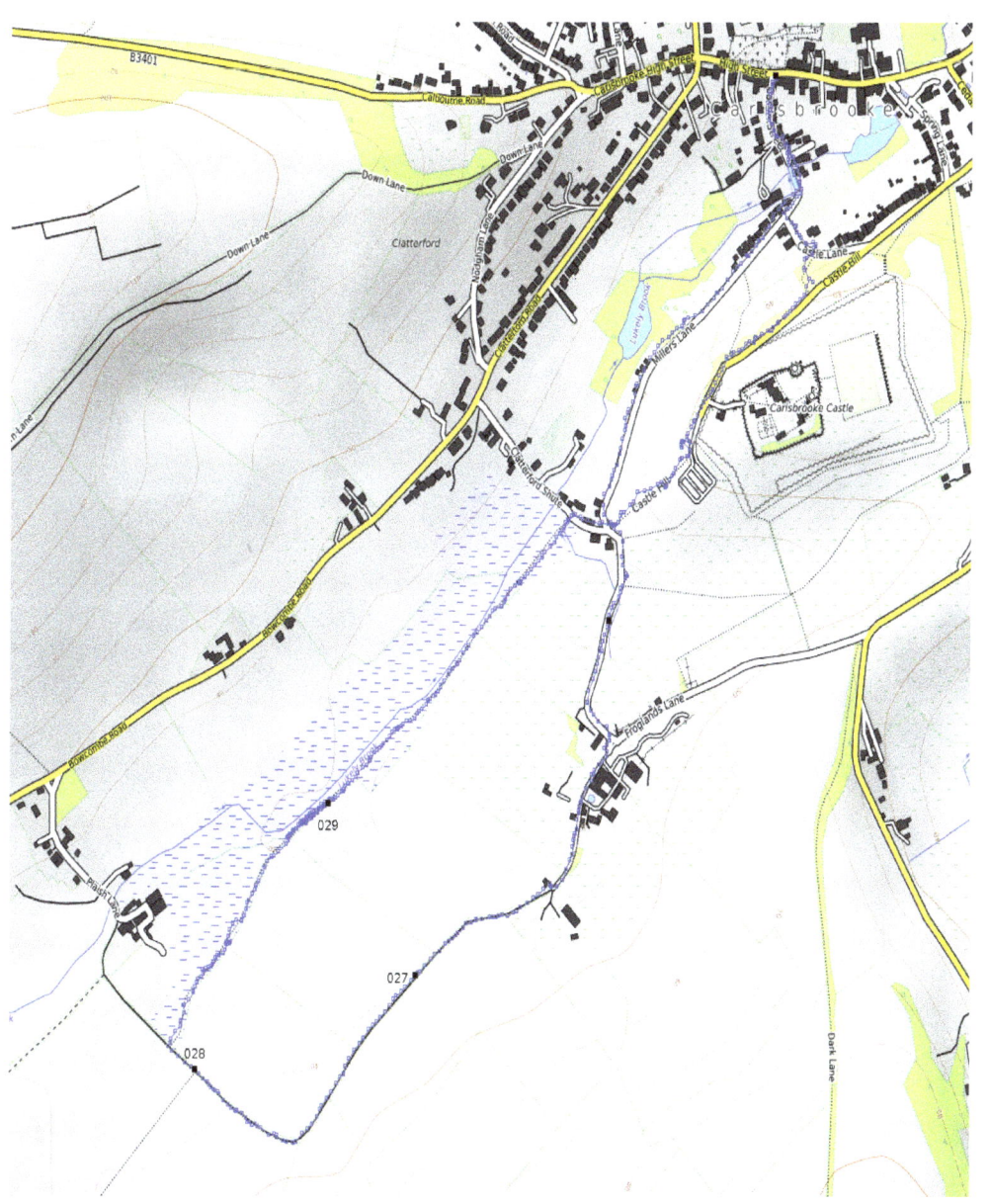

Not recommended in, or after, a rainy period; this route can be very muddy, boots essential. As described this walk is a 'figure of eight' rather than a circular one. There are options to extend or shorten it.

Make a start from Castle Street in Carisbrooke; almost opposite the Church. As the name implies this is the ancient and direct route between the Castle and the Priory Church of St. Mary's and is full of interest in itself. Note especially the Wayfarers Fellowship Evangelical Church, which started out in 1859 as a Primitive Methodist Chapel; also the house next door which has grown in height!

Opposite the chapel is the entrance to a mill which ceased working in 1939 and is now residential. I have always known it as Kent's Mill. Kent was, apparently, one of the later owners, but over the centuries it has had various other names; Kings Mill, Castle Mill, Longs Mill, and inevitably Carisbrooke Mill; strangely Priory Mill was the next one, toward Newport. (Now the waterworks.)

Take the footpath alongside the ford, and turn right along Miller's Lane ignoring turnings to the Castle on the left. At the end of the bungalows, on the right, note the very ancient roller now languishing on the grass outside someone's gate; what could it have been used for? Continuing along the lane we have a laid hedge above the bank on the left and the ground on the right falls gently away to the flood plain of the Lukely which in the past would have been a well-managed water system damming the brook to form the very large millpond. In amongst the greenery on the other side of the main stream, unseen now, is the long, concrete leat toward the mill which fed the water onto the water wheel.

There are one or two more houses along the lane but there is an 'away from it all' atmosphere to be enjoyed in these lanes

At the end of the lane bear left but do not take the footpath N88 to the immediate left which goes up to the Castle; continue on the metalled surface over a slight rise and down to another ford, certainly one of the prettiest on the Island. The brook here is an unnamed tributary of the Lukely (but I have heard it referred to as Frosthills Stream and the ford as Fourways, perhaps a reference to the junctions either side) which issues from springs just across the field under Froglands Lane in a little area called Frosthills. Somewhere hereabouts in the field to the left there was a small paper mill making use of the water as it dropped down to the level of the ford; listen for the little waterfall, where the sluices used to be, on your left as you carry on up the lane. There is a small millpond above this, behind the hedge.

When you reach the junction with Froglands Lane turn right. This junction is probably as close as one can get to an identification of the place known as Frosthills (Known in 1150 as Vorsteswelle, the name undoubtedly predating Froglands or, in 1395, Froglane).

You soon come upon Froglands Farm. Have a good look at the first barn; end on to the lane. Can you see the brick inscribed 'Jane Stark 1792'. A little bit of farming history for someone to follow up. No longer a working farm in the normal sense, it has more to do with horses, it has, nevertheless, a wonderfully traditional farmyard complete with duck pond and surrounding buildings, quite a picture.

The lane now becomes more of a farm track between, surprisingly, quite robust and tall elm saplings; when, I wonder, will they fall prey to the Dutch disease? The track rises slowly for a short distance up to a point where a bridleway (N101) starts; this bears slightly to the right. From here on, in wet seasons, the track quickly becomes very muddy due mainly to horse 'traffic'. You soon come to the end of the elm saplings and the track takes a slow curve to the left.

As the way straightens out, there is an improved view of the surrounding high-ground. Alvington Down on the right with its No 3 reservoir near the top, then it merges into Bowcombe Down followed by the Downs of Idlecombe, Rowborough, and in the distance Cheverton Down. Above and beyond Bowcombe Down is the north-east corner of Brighstone Forest and the Rowridge TV mast. To the front and left is Dukem Down which merges into Garstons Down and behind that is the Chillerton TV mast.

**It is between this curve in the track to the left and the next field boundary on the left that PLUTO crossed this path. There are markers here; one of them a good one.**

You are now in open farmland with plenty of interest in terms of birds and plants; the last time I went this way there were no fewer than four buzzards soaring and hunting above these fields.

At the end of this stretch of path it curves to the right but before continuing, take a look into the field gateway on the left (If possible; sometimes it's grown over) and admire some unspoilt Island farmland scenery. Unspoilt but not untouched, note the low cliff face in the corner diagonally to the left which shows how close the underlying rock is in this area.

I am told that stone was quarried here for use in building parts of Carisbrooke Castle; I wonder if it was the Normans or Elizabethans or someone in between? The field is apparently called Rancombe; was it named before or after the, now demolished, farmstead of the same name near Shorwell? There was probably some connection.

From the turn, the path goes toward Plaish. Slightly downhill initially, between mature hedges of hazel with a few very contorted ash trees. Underfoot the path has either been heavily cobbled or it has worn down to bedrock; whichever it is it holds water rather well! After about two hundred yards a footpath, N205, takes off across the fields to the left; this could be used to extend the walk by about a mile if desired (see EXTENSION at end of the chapter). However, go a few more yards down the track and turn right over a stile, this is the start of the, now diverted, N104.

** The stile takes you into a region of natural water meadows. This path used to commence near the old sheep wash at Plaish itself; that is the farm and houses which you can now see to the left, but it has fairly recently been diverted and now cuts out the initial part of the path alongside the mainstream of The Lukely. (Many of the path diversion of recent years have deprived the public of much that is of interest; especially with respect to farms and buildings).

The marshy ground, and sometimes quite a pond, immediately to the left of the path at the start, is one of the springs which form the first little tributaries of The Lukely which itself starts further up the valley above Bowcombe. I must here recommend Bill Shepard's book 'Newport Remembered', the early pages of which give a fascinating description on the tapping of these waters for Newport's water supply including the tunnelling that took place. It's difficult to imagine men toiling away, actually under these very wet meadows.

Continue with the hedge on the right. There is a wealth of bird life in these meadows especially in summer with swallows skimming the water. Further away to the left on the main Newport to Shorwell road is first, Goldings and then Bowcombe Chapel. It was what was described as a 'cloudburst' over this valley in the early hours of the 1st October 1960, coupled with a very high tide, which caused the severe floods in Newport on that day. For a few hours Whitcombe Road was the only way into town. (Except for the railways from Ryde and Cowes).

Bowcombe may seem a small place today but this whole area had once been known by that name and the Bowcombe Hundred was one of the three Hundreds that the island was split into. In fact, it seems that what we now know as Carisbrooke church was once Bowcombe church and it may even have stood further up the valley!

Just before the hedge into the next meadow is reached the small stream disappears through the hedge where it joins The Lukely, now coming in from the left.

It is between the stile into the next meadow and the electricity supply pole a little further on that PLUTO crossed the path. There seems to be no trace of a marker here, although one was recorded in poor state shortly after the Millennium. Did they bury the pipe under the stream to the left? I wonder?

The path now runs alongside The Lukely itself with the castle looming to the right front and Carisbrooke church visible, slightly to the left, ahead. In the fields the other side of the stream, lie some Roman remains; Bowcombe villa. After discovery in the 1850s it was reburied and it seems that it has only been disturbed once since then. There are said to be three Roman villas in the Carisbrooke area; the one mentioned above, usually referred to as the one at Clatterford, one further up the valley, known as the Bowcombe one, and one in the vicarage garden. I've not seen any location reference for that said to be at Bowcombe

Some way further on there is what appears to be a small marl pit to the right, which seems, for some reason, to be channelled down to the stream! After this, the path is very close to the stream and can be tricky going until the next stile is reached.

This is now the last of the meadows, very flat and much more like a purpose made water meadow for wintering livestock. This could be the meadow called 'Court Mead' There is a comparatively recent and quite elaborate sheep wash a little to the left made from concrete blocks. Recent! well early 20c, possibly replacing an earlier one. The path now approaches the lane known as Clatterford Shute; the large house, three up from the ford, was once the home of Col. Brannon, proprietor of the County Press; more recently, for some years, it was a pub, the Shute inn, but is now a private residence again.

With the stile onto the lane in sight another tributary emerges from the garden to the right; this is the stream from the spring at Frosthills whose ford you passed earlier in the walk. It now sweeps between the path and the stile before joining The Lukely at the ford on the lane to the left; hence a footbridge has now to be crossed to the stile. Not long ago it was a few slippery stepping stones!

The Clatter-ford to the left gives its name to the whole area. Apparently it gets its name from the Old English 'clater' alleged to mean loose stones or pebbles; or did it just mean that going through the ford caused a clatter. I wonder which came first? All academic now as it has a concrete bottom. But it does bring to mind a verse of 'The Brook' by Tennyson: -

> *I chatter, chatter, as I flow*
>
> *To join the brimming river,*
>
> *For men may come and men may go,*
>
> *But I go on for ever.*

In the lane turn right. In a short distance Millers Lane joins on the left; which was your outward route. Rather than retrace your steps take the path N88 next door on the left leading up to the castle; a very ancient track this, as old or perhaps even older than the castle. It is well worth stopping occasionally on the way up, to turn and admire the view over the Bowcombe Valley. Near the top to the right is a convenient gateway for a rest and to soak in the views over Garstons and Dukem Downs.

The path emerges through the car park on the south-west corner of the castle. I can add nothing to what has been written about Carisbrooke Castle except to throw in a thought on its genealogical royal connections which I think needs more delving into and, if clarified, made more of by the island!

It would appear that Alfred the Great's mother was a royal Jute from Carisbrooke and one of Alfred's daughters (she who inherited Wellow, IOW) married Baldwin of Flanders who was an ancestor of Matilda, wife of William Duke of Normandy (The Conqueror). And William's grandfather was the brother of Edward the Confessor's wife!! So perhaps William wasn't quite the outsider that he is often popularly considered.

Histories aside, continue the walk across the front of the castle. To the left, the parking area is a magnificent viewing platform. Alvington Down is seen to the left front, with its number three reservoir prominent on top. Below it, the whole of the south western outskirts of Carisbrooke are spread out in a variety of housing styles; much of it, north of the road, built in the former grounds of the large Italianate house in the centre.

There are two paths down to Carisbrooke from here, the first, N191 starts at the corner of the outer defences between the old style traffic lights and descends to the left between the trees. This path crosses and comes out into Millers Lane; it can be difficult when wet or frosty.

The other path, which is preferable, starts a few hundred yards down the road, soon after reaching the avenue of holm oaks. Trees are quite a feature this side of the castle, the beeches by the first path, the cedar of Lebanon up near the keep and this second path N192 goes down steps on the left amongst some more wonderfully tall and stately beeches.

The steps bring you out into Castle Lane and turning left into this lane will take you the short distance to the end of Millers Lane by Castle Street ford and hence by the ford and back up Castle Street to your starting point.

### *EXTENSION*

This walk may be extended if you wish by taking that footpath, the N205, to the south-west at the farthest point of your walk. It goes straight across the fields heading towards Garston's Down. A lovely walk, on a good day.

Having crossed the fields, you come to a track running from right to left. This was (is?) the unmade road from Bowcombe to Gatcombe. Turn right onto this chalk track toward Bowcombe Manor Farm; this section is designated the N144. Just before reaching the farm you need to turn right into another, this time quite sheltered track, the N102.

This takes you back toward Plaish. It is a wide track with its high hedges and is typical of so much of this valley; in that it still abounds with nature and, depending on farming activity, can be unbelievably tranquil.

The turn to the left at the end would take you to Plaish Farm and main road but keep on the track going around to the right; you will now be back on the N101. In about two hundred yards you will come to that stile, now on the left, at the start of the diverted N104; that footpath across the water meadows. Continue from there as above **

# Walk 5

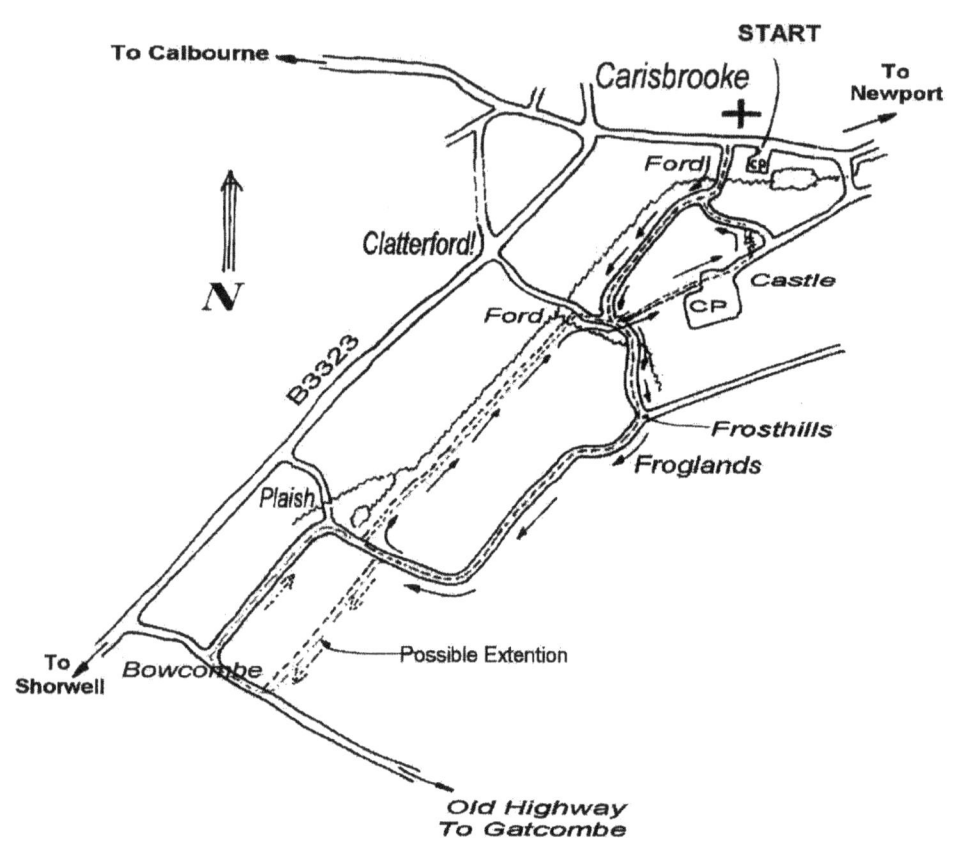

*Not to scale*

# 6. APSE AND DOWNS

*(Between Whitely Bank and Shanklin)*     *(Approx. 3¼ miles.)*

I found great difficulty in deciding where to start this walk with safety. The road from Whitely Bank to Shanklin is not one I would walk unless the car broke down! There is a bus stop, and room to park a car, at the junction of Apse Lane but to access a footpath you would need to walk east or west on the main road or to Apse Manor. This was not always the case; there used to be a footpath (the continuation of SS13) going across to the old railway opposite Apse Lane; this badly needs reinstating.

However, as Victoria Avenue, Shanklin, has pavements and the remaining section of SS13 is usable its best to start from there. Coming from Shanklin, footpath SS13 starts with a stile alongside a field gate at the end of Victoria Avenue just as you start to turn to the right toward Cliff Bridge and just after the drive to Cliff Farm on the left. This is a short path through two small paddocks. At the right time of the year there are orchids in the first one. The path ends with a 3rd stile and little bridge over a deep ditch and a 'T' junction with footpath NC39.

Turn left onto the path which now heads for the downs with wire fence on the right and the deep ditch and bank on the left which, from mediaeval times, had been the Parish boundary between Newchurch and Brading and the manorial boundary between the manors of Shanklin and Apse. This is a lovely path with vistas of the fields running up to St. Martin's Down on the right and Shanklin Down on the left. The path aims broadly at the saddle between the two.

All the way up on the left are horse paddocks, which now abound on the Island, taking up what had mainly been, and could again be, good farming land; even though in the late 19c the first of these paddocks contained brickworks and kilns.

About one hundred and twenty yards along this path a stile is reached, alongside a hedgerow coming in from the right. **It is at about this point that PLUTO crossed the path.** I say 'about this point', deviating from my usual comment of between two points, because, although the lack of a marker is not unusual, I believe this crossing has deeper secrets and a piece of PLUTO pipeline may still be below!

Just over this stile another stile gives access to the field on the right which is the route of NC39. Take this if you want to amble in the field or look for mushrooms, it rejoins your route further up by forking left in the second field as the NC39a (the only fork visible is on the map!). The right 'fork' at that point is the continuation of NC39 along the bottom of the inland cliff. The main path, perversely, becomes NC39b; confused? I was.

The main path by the hedge, although it appears old and indeed may be, was not a formal footpath until the late 19c; hence its later designation; the one through the field was obviously designated earlier (Beware; the paths in the fields are not signed, nor obvious).

So, continue on the main path, the NC39b. That pronounced ditch on the left sometimes appears to be more of a mini ravine. There are some fine oak trees in that field to the right, also to be seen in that field is a good example of how much the land here, given the right conditions, can slip, even in a seemingly stable pasture. Some of those fence posts on the right are a great testament to the durability of wood when used rustically.

This is a great path in early autumn for sloes, blackberries and hazelnuts. If tempted, you could take a long time going up this path.

You soon come to a grove of oaks on the right, this time covering a knoll in this increasingly disturbed landscape. There is little to view on the left, on this part of the path, due to the high hedge of saplings and older hawthorns and hazels giving way to a copse.

A slow flight of rustic steps now takes you up to a stile and field boundary. At the stile, without its arms is one of those fine old footpath signposts cast in Newport by Wheeler and Hurst; rather more durable than today's offerings made of steel pipe! Along the field boundary to the right can be seen the stile for the footpath mentioned earlier, coming up through the fields.

The path now continues alongside that left-hand boundary unconfined to the right in increasingly rugged pasture. The view to the left is better now and on that horizon, directly to the left, can be seen a row of trees which is the western edge of Hungerberry Copse. This is the copse which was used to conceal the huge header tank for the PLUTO pumps on Shanklin and Sandown waterfronts. And, of course, the destination of the pipeline coming across the Island from Thorness Bay.

As you walk up through this field you come upon a concrete water trough for the cattle fed by natural springs and land drains. An example of sound ecological practice but by no means new. Unfortunately, in the last few years it has more often been virtually dry; we so rarely now have the situation so wonderfully put into verse by John Greenleaf Whittier which, being now in the hills, gives me an excuse to quote what is amongst my favourite pieces of poetry:

> 'For weeks the clouds had raked the hills
> And vexed the vales with raining
> All the woods were sad with mists
> And all the brooks complaining'
>
> From 'Among the hills'

As you climb steadily up through the field the rock face of the inner cliff rears up ahead; your path lies up through it. Just before you reach the next stile the 'invisible' NC39a, coming from the NC39 in the fields, joins from the right. When you reach the stile a look back, and a rest, are worthwhile.

Much of the north east of the Island and beyond is laid out before you. Depending on visibility the view to the far right is of the Sussex coast to Selsey Bill and beyond; Shanklin, Sandown and Culver cliff on the Island are prominent in the foreground. Further round to the left is the white sea-mark on Ashey Down and then the downs of Mersley, Arreton and St. Georges. More to the left are the Island's western downs with their two TV masts and even a glimpse of the white cliffs below Tennyson Down – these become more visible shortly.

Near at hand, along the fence is yet another stile, this time for NC39 going away westward. Over the stile hug that left hand hedge for a while; the path takes you up through some rather scrubby hawthorns and when you emerge from them some steps will be seen, slightly to the right front, going upwards of course. At about this point a little subsidiary link path goes off to the right to join NC39 going along the bottom of the cliff; it is not marked nor does it seem to have a number!

There used to be three cottages along that path through the fields at the bottom of the cliff; it probably explains why that path was formalised earlier. The place was known as Apse Reach. It would appear that it was abandoned around 1900.

Proceed up the steps, which gradually become more organised, and you reach the start of the steps up through the cliff which is presently fitted with handrails. Over the years, steps and rails have come and gone, due to wear, tear and weather, but the present ones seem rather more permanent. Having attained additional height, a backward glance reveals more of the Island's centre; the glass houses in the Arreton valley, the surprising amount of new housing beyond Apse Heath and more of the mainland. The tip of Portsmouth's Spinnaker Tower, the chalk quarries behind Portsmouth, the coastline at Lee-on-the-Solent, Fawley Refinery and The New Forest all emerge beyond the Island.

This cleft up through the cliff is no doubt partly natural and an important feature of the parish boundaries. Did the fact that there was a way through the cliff here dictate the boundary or was it the other way round? It was only consolidated as a formal path at some time in the years between 1885 and 1908. Before that it would certainly have made beating the bounds interesting.

The cliff itself is mainly the Upper Greensand, a geological definition which includes various types of stone given common names i.e. freestone, frestone, rubstone, chert, sandstone, etc. All used on the Island in the past for buildings. The rubstone so named for its use in rubbing hearths and doorsteps.

A gate at the top takes you into a corner of Shanklin parish and a beautiful high meadow stretching away to the left. (The footpath in that direction, the SS10, would take you to St Blasius old church at Shanklin.)

However, turn to the right here and almost immediately another gate takes you into Ventnor parish and a continuation of the meadow where your path lies ahead along the tree-lined top of the inner cliff. (Ignore the path uphill to the left after the gate that is the SS12 to Ventnor).

The cliff top here marks the boundary between Newchurch and this much newer parish of Ventnor, so the path now becomes the V44 Both this path and that path back to Shanklin are part of what is also now known as The Worsley Trail.

The path keeps a few yards into the meadow from the cliff top, which cannot really be seen. It is fenced to keep livestock from the edge and it is unwise to explore too closely, there have been fatalities!

There are no particular points of interest along here, it's just a very enjoyable walk and nature abounds. The higher ground to the left is the saddle between the Shanklin and St. Martin's Downs. The path continuously curves around to the right. Mature ash trees line the cliff top further on and soon after that another muddy patch takes you onto the open shoulder of St. Martin's Down. The path now leaves the cliff top and hedgerow to the right and continues virtually straight across the slope.

The views now, having gained height, are even more extensive to both the east with more of the Sussex coast exposed and Bembridge Down with its fort in the foreground and also to the north-west, taking in even more of the Island.

Soon after the high point in the path, it is joined by another footpath, the V45, coming down from the left around the side of a massive marl pit which looks very much like an Iron Age earthwork. (The two paths join to become the V46; but only for a short distance)

Marl is something of a general term for any calcareous (chalky) material, sometimes rock like, often mixed with light clays, which was much used in the past for dressing heavy soils; hence pits all over the Island where suitable material was to be found. The burning of rock to produce lime was a further development of land dressing; the Island has many remains of lime kilns. (Some brick earths are also referred to as marls)

The path now starts to drop toward a fairly obvious gateway in that right-hand hedgerow. Ignore the path to the left just before the gate, that is the continuation of the Worsley Trail, go through the gate where a gentler, diagonal, cutting takes you down through the, now less obvious, inland cliff. Ignore the gate set back to the right; ignore also a path to the left, into the trees about thirty yards down.

Carry on down the sunken track. There are some great trees here, sheltered from the prevailing winds of the south-west, most having even escaped the hurricane of 1987 from the south. Obviously a very ancient track to the downs this, as witnessed by the depth of its banks.

At the bottom of the cutting some six footpaths all come together but before you get to the path signposts, (there are two a few yards apart) turn right into the first gateway on that side and go through a bridle gate into a field. There is another signpost here showing the NC39 going back under the cliff and the one you need to take, the NC30 to 'Apse Heath and Main Road' (Apse Manor would be more accurate). This path continues on down through the field alongside the fence to the right.

Not an old path this, it only dates from the late 1800s, but a lovely one nevertheless, with the shining sea away to the right and on it the black mark of The Nab Tower well out across Sandown Bay.

Halfway down the field, well hidden under a clump of hawthorns is another water trough (actually an old cast iron bath) making use of a natural spring. This used to be the start of small stream as can be seen from the vegetation along its route but it has been woefully dry of late.

The path drops into a small grove with a massive oak to the left over that stream, which may have gathered a little more water from the fields by now. This is a delightful spot for a pause, perhaps for nammet!

Through another bridle gate the path, after a few yards, joins a much older farm track with what is now quite a little ravine, fed by more springs, on the right. The track is little used by vehicles until the next gateway is reached which leads to a Dutch barn on the right. However, your route continues down the main track; now gravelled.

Some cottages, of 1902, are reached on the left. It is around here that this track was crossed by one of the old bridleways from Shanklin to Whiteley Bank before the main road was opened in 1885. Most of the east west paths and bridleways in this area seem to have been extinguished once the road was opened

**It is somewhere between the end of these cottages and the other side of the bridge ahead that PLUTO must have crossed the path.**

Unusually I say 'must have' because my sleuthing has failed to reveal the crossing point which also had here, the additional problem of crossing that now very deep ravine to the right. Common sense and my intuition suggest that the engineers would have made use of the railway bridge to carry the pipeline over both track and ravine. Also, at some point nearby, it must have been laid under the rails. But there do not appear to be any markers; at least not now!

Take great care here. After the cottages, there are a few places where the edge of the track, on the right, is the lip of the ravine. Peering into the hedge could be hazardous; it's quite a drop.

Just beyond the bridge some steps on the right lead up onto the old railway track. Having now finished with PLUTO on this walk you could cut it short here by turning left along the track at the top and taking the next footpath along it to the right (the NC39 again) to regain that little path across the paddocks which you started from.

However, to make a somewhat longer walk of it, without using the road, continue the short distance down the track to the main road. This is that main Shanklin to Godshill road, which opened in 1885. Your path, thankfully, lies directly opposite and is a continuation of the track you have been on; it is a direct route from the downs to Apse Manor and predates the main road by many centuries. The path is now signposted as the NC30a to 'Apse Manor Farm and Apse Lane'.

This side of the main road the ravine and its little stream veers off to the right, to the side of Apse Lane. The stream eventually flows through the back of the manor grounds up ahead. There are now more horse paddocks to the right. At the end of the path bear right on the field entrance track; this takes you through a small copse onto Apse Lane.

Across the lane is the ancient Apse Manor house, parts of which date back to the early 17c. One room famed for its oak ceiling. In recent centuries it has been used as a farmhouse and many of the farming families of the Island have been owners or tenants.

It is now believed to be the manor written into the Domesday book of 1086 as 'Abla', in which case, it was held by King William as part of his manor of Niton. It was already spelt Apse by 1100 and any distortion of the name after that is just down to bad spelling; which is odd if, as is said, it has to do with aspen trees! With other properties on the Island, The Cannons of Christchurch once held it as a grange.

Turn left into the lane, but take extreme care for the first fifty yards, as it is a blind bend and narrow. As the road straightens the surrounding buildings can be better appreciated. On the left are the conversions of an old forge and associated sheds. Ignore the footpath (NC32) going off to the left; that crosses fields which used to be, in the 1920s, the Island's first aerodrome.

The real beauty of the old manorial farmstead lies to the right. A wonderful range of brick built barns and cart sheds with classical archways, haylofts etc. I would have preferred it to have been kept as an intriguing collection of farm buildings. However, 'progress' has demanded that it becomes dwellings; at least the conversions have been done quite tastefully. Note the old rotary pump down in the valley at the end of the main buildings over what must have been the main well.

At the end of the straight, as the lane turns up to the left, tucked in the very corner, after the entrance back to the barn on the right, beneath some 'fir' trees, you will find an optional stile (look behind the hedge). The footpath is not marked but it is the NC37b into America Woods. Those 'fir' trees you duck under are Monterrey cypresses. Some rough steps take you down to the valley floor where the path bears round to the left; the area to the right is gradually being taken into care as part of the gardens for the manor complex, and now houses alpacas. The path now goes across the valley towards the right hand boundary and follows that boundary around to the right.

A footbridge takes you over the stream, the same one that you have followed all the way down from St. Martins Down, and into a bracken lined path for about one hundred and twenty yards, via a stile to another stile into the woods proper. (That stream becomes Scotchell's Brook)

America Woods, the name is intriguing, some old maps label the cottage in the woods 'America Cottage' so whether it was the cottage or the woods first named may not now be known. The name dates back well over two hundred years. It has been suggested that it may be connected with oaks and shipbuilding.

Over the stile there is a short uphill slope with a 'T' junction at the top. Turn right here; this is not the formal footpath, which goes left, but a shortcut to the main path through the wood. After a few yards a sunken path amidst the sweet chestnut trees can be seen, with felled logs on the left. This is disused but keep it to your left.

The path takes you up through the bracken to the main path, the NC37a. There is no signpost.

Turn right here along this very sandy path, (Apse sandpit is not far away to the right); there are many twists and turns but ignore all side turnings, the main path is fairly obvious. The path's general direction is always south.

There is a wonderful variety of trees in this ancient woodland, massive oaks, taller than usual silver birches, sweet chestnuts, lots of mature hazel, holly and many more. The woods are under the care of The Woodland Trust.

It seems quite a long way back through the woods to the main road but it is a glorious walk in any season. Eventually, provided you have resisted the temptation to explore those side turnings, you will drop down and cross the little stream running through the bottom of the woods. Keep left after the bridge and it's a short climb up to the main road. Fortunately, it is possible to go straight across the main road again here; this time to footpath NC39 signed to 'St. Martin's Down and Old Railway Track'. This starts as a private road with bungalows to the right and a heavily wooded ravine to the left; it is only a short distance to the old railway which is now a public path and cycle way between Shanklin and Wroxall.

Cross the old railway to continue on NC39, now signed to 'Shanklin Down'! It is only about one hundred yards now to the junction with SS13 on the left, which goes over the footbridge and stile to take you through the little paddocks back to your starting point at the end of Victoria Avenue.

**Walk 6**

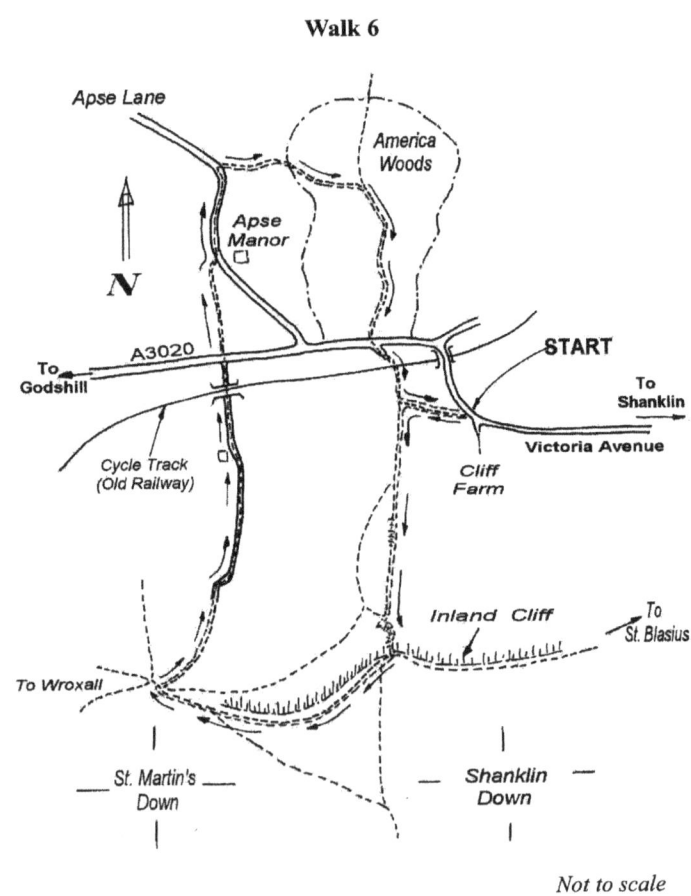

*Not to scale*

# 7. NEW PARK AND OLD MANOR

*(Ambling west of Newport)*   *(About 4½ miles)*

West of Newport there is a square of what is now farmland, enclosed by four roads; Forest Road to the north, Gunville Road to the east, Calbourne Road to the south, and Betty Haunt Lane to the west. There are a number of footpaths in this square with nine access points, five of them being farm entrance lanes, the other four are just footpaths. The area was, in medieval times, part deer park and part forest, all to do with hunting. Very broadly the area to the west of Betty Haunt Lane was 'Old Park' and this area, to the east, was 'New Park'. Over the centuries it has been converted into farmland and makes for interesting walking with a number of variations. It is rather flat, low lying country but it makes a change from the Islands usual more dramatic and panoramic routes; although the walks are short there are possible extensions to the north and west.

The commencement of this walk is via the Footpath N83 that starts from the main road through Gunville. This area was alleged to have been named from the French, Gonneville and to have been on the route of the ancient tin road across the Island from Gurnard to Puckaster. The 'Tin Road' theory is now considered very unlikely due to the alleged fording of the Solent and its geological versus historical time scale. However, if it is discounted there is a lot of 'evidence' to be otherwise explained.

The entrance to the footpath, just south of a road called 'Alvington Manor View', is co-located with that of a fishing pond; it has a large white sign saying 'Gunville Lake' together with a finger post for the footpath. Forty yards from the entrance the footpath veers off to the left, while the main track to the 'lake' continues to the right. After a short woody section, you come into more open countryside with Alvington Down away to the left and a little later the 'lake' can be seen through the bushes to the right.

The 'lake' has formed in pits that had been dug to provide clay for brick making. There was a sizeable factory here, with its own railway siding; for a short time the factory also made high quality pottery. Originally it was owned by the Prichett family but was bought in 1923 by S E Saunders the boat and aircraft manufacturer, of East Cowes. Considering its central position and its rail connection it is surprising that it was so unprofitable as to have to be closed down in 1937. Apart from the clay pits the factory site has been obliterated by modern development. Intimate knowledge is required to work out where the railway went under the road and through the area.

Soon after the 'lake' is lost to view, a gate brings you to the junction with the N85 which makes use of a lane coming from the end of Alvington Road, Carisbrooke (although signed to 'Gunville' from here). This lane is one of the entrances to Alvington Manor. Turn right onto this, now substantial, track which quickly leads into the manor complex of cottages and farm buildings. More of Alvington later.

Pass the dwellings on the right and, opposite the centre of the barns on the left, turn down the track on the right leading downhill away from the farm. There is no finger post but there has been a yellow arrow. This is FP N151. On this walk especially I have only used FP numbers where they are fairly straight forward. Note that some of the FP numbers on guide maps for this area are confusing. This is particularly the case from Green Park back to Alvington

When you come to the first field boundary either side of this track you cross the tree-lined remains of the, long closed, Newport to Freshwater Railway. Not much to give it away now, only perhaps the quality of the iron gate pushed into the hedge on the other side, and the raised track bed. The line opened in 1889 and closed in 1953. The route was bought by the County Council for about £3000 but then sold off to the various landowners. If only the council had kept this permanent way, what a wonderful route it would have been for cycling and walking through the lowlands of western Wight.

As this land is low lying there is little to be said of vistas; however, Parkhurst Forest can be seen ahead and to the right is the urban sprawl of Gunville Backs, now reclaimed and renamed Gunville West.

At the bottom of the next fields which the track goes through, the track winds around several twists – there are old stiles in the hedges around here as the route has been altered. Note the line of willows across the field, they mark the course of the brook which goes through Gunville 'Harbour' and joins the Lukley in Newport. In medieval times the brook was the southern boundary of the forest. The Normans called it Esthbroc which is a puzzle because although it flowed east it was to the west!

In more recent years it has been called Sommers Brook or more often now, simply, Gunville Brook. (The Lukley was called Suthbroc presumably because it was to the south).

Keep left on the N67 which passes into a field. Follow the hedgerow and brook, keeping it to your left, past another field entrance still with the hedge on your left, you approach Poleclose Farm and emerge onto the farm end of Poleclose Lane (N68) with its line of oak trees. The path used to go straight ahead through the farm. It has now been diverted so it is necessary to turn right here, toward the main road, but about one hundred yards down the lane turn left into a track, this is the N69, which goes across the front of the cottages and farm; albeit now some distance away. The names of farms in this area have been very confused over the years, possibly due to families moving around. Poleclose, on a map of 1770, is shown as Paulchose (!) and in 1781 was referred to as 'or Cockleton' but later this name was also given to Cooks Farm further on in the walk.

There is a newly formed pond on the left of this gravel track. As you come level with the farm and where the track turns slightly left, uphill toward some of the farm's barns, the footpath, not very distinct here with slightly misleading arrows, leaves the track and bears right through a presently unused piece of ground. After a short distance, when clear of the end of the barns to the left, the path emerges into an often boggy meadow. The original path ran nearer the hedgerow away to the left but, due to the diversion, you now need to go virtually straight across the meadow. Aim for the prominent oak tree ahead, but beware that in a wet season that tree will be at the edge of a small pond. On reaching the vicinity of that tree, leave it to the left and follow that hedgerow around to the right, with the 'hedge' on the left. This area appears very 'set aside' at the time of writing but, thankfully, this adds greatly to its attraction; it's very natural and a haven for wildlife and plants.

Following the path around the edge of the meadow you will come to a gap with arrows; just past this the path crosses a crude footbridge over a ditch, taking you to the right, though still keeping the hedge to your left. After wriggling around this seemingly forgotten corner, (Old maps show buildings in this area) for only a few yards, a stile takes you onto another farm lane. This lane, off Forest Road, the main road between Newport and Yarmouth, is the access lane for Green Park Farm and Cooks Farm; it also connects with Reads Farm.

Opposite the stile across the lane is Park Green Farm itself, once an establishment of the Biles family. In the not too distant past the farm was one of the few abattoirs left on the Island but, due to EEC regulations generally preventing slaughtering on the Island, the farm is little used and the farmhouse was demolished some years ago. (These regulations are quite a problem for Island farmers and others with livestock).

After crossing the stile turn left down the lane (N71) alongside the farm. The farmland either side of the lane now has a somewhat more used look about it. Ignore the footpath (N78), with an un-numbered fingerpost, to the left and continue. The lane meanders slightly at this point but for a short distance it runs parallel to the line of PLUTO in the field to the right. **Just before a square area to the left and as the lane bears slightly right PLUTO crosses the path, but at a very acute angle.** There being no hedge to the lane there are no markers; but there was a marker to be found in the corner of the hedge to the left front, bordering that area, which is fenced from the lane. In times past it was the site of Heath Cottage, which explains why it is as it is.

The lane continues to Cooks Farm; known in 1862, and later, as Cockleton. At the entrance to the farm the path bears away to the right. There is a gate and stile before yet another gate and stile taking you into the yards of Reads Farm only a short distance away. After the second gate you would once have immediately recrossed the Newport to Freshwater railway, which is still quite evident. A small field away to the right there used to be a small holding called Green Park, but not to be confused with Park Green above!

Reads Farm is presently owned by a family called Reed, indeed the map of 1862 shows it as Reeds. It is quite an extensive and obviously very productive farm with, at the time of writing, a large herd of Friesians; as shown by the weathercock. The path goes straight through the farm, a nice change from being diverted away from anything interesting; further on, the drive is lined with old saddle stones. The delightful, and surprisingly small, farmhouse can be seen to the right as you leave the farm buildings behind. Follow the drive, which turns to the left as you leave the farm. Just ignore the footpaths (N73 & N148) which take off into the fields ahead and to the right at that corner of the drive.

You now climb slightly between hawthorn hedges with a little more of a view; Alvington Down to the right and Parkhurst Forest to the left. As you come to a small grove of oak and ash to the left the drive turns away up to the right. Just past the last tree on the left, where the drive finally turns and past the field gate on the left, is a disused stile in the corner ahead. That gate on the left is a good place for a pause to soak up the surrounding area, it overlooks the new house at Cooks Farm and the forest beyond – and, unfortunately, the recycling plant.

The stile in the corner takes you into a field with a hedgerow on the left. This route was the original entrance to Read's Farm before the drive on the right was constructed. The chimneys of Alvington Manor can now be seen peeping above the trees to the right front and slightly further away to the right front are the trees on the hanging concealing the main Newport to Calbourne road. **There are two oak trees to the left as you go down the side of this field and at some point between these two trees, PLUTO crossed the path**. There is most of a marker here, at an angle.

At the end of the field you enter a reasonably unspoilt area. The path, now a farm track, winds around to the right and left between waste ground; that on the right was the site of Sheepwash Cottage in the early part of the 1900s. These areas and the trees all contribute to good habitats for wildlife. Ignore the track to the right and the FP to the left, which is the other end of N78 which was passed earlier. Continue on the main track which is now something of a causeway between two very deep, tree lined, ditches; one of which was, no doubt, the watercourse dammed to form the sheepwash. After wet weather these look more like large ponds.

After the ponds you approach the ancient manor of Alvington and I would like to dwell on this a little longer than usual. I don't feel that Alvington gets the attention it deserves in various histories and guide books. Although close to Newport its location is secluded, it hasn't tourist appeal, and over the centuries farm building have somewhat eclipsed it; but it is important. To put it in perspective, Carisbrooke is not mentioned in the Domesday Book; the only reference to the castle is under the Manor of Alvington (Alwinestune) in that it has a castle standing on one virgate of its land! Obviously considered a bit of a nuisance in that it restricted the agricultural value of the manor.

The present house is only a little over three hundred years old and hasn't got the same attraction as say, Arreton or Wolverton, but historically, as a manorial holding, it is just as, if not more, important. Before the conquest it was held by someone called Dunna and at Domesday by King William. After this it was held by many of the renowned Island families. For about 260 years from 1100 it was in the hands of the St. Martins. The St. Martin heiress married a Popham and three generations later the Popham heiress married a Wadham who, after three more generations, produced a Captain of the Island; Sir Nicholas Wadham (1495 –1517). The Captain actually sold the manor, probably the first recorded sale, to the Harveys who in the 17c brought the manor into the Leigh family, by marriage. As the feudal system declined it became more often referred to as, merely, Alvington Farm. Rushing through a thousand years of history brings to mind two lines from Kipling verse : -

'His dead are in the churchyard – thirty generations laid.
Their names were old in history when Domesday Book was made'
From 'The Land' by Rudyard Kipling

Unfortunately, only rare glimpses are possible of the house itself. The farm has become rather a sprawl but some of the older buildings have recently been converted to holiday accommodation. Ignoring the entrance drive to the right continue to the left of the converted barns where further ponds are being developed away to the left close to the old railway line backed by a line of poplars. As you come to the more modern barns you will rejoin your outward path. Continue past the cottages, on your right, and take the N83 back to Gunville or stay on the track, now the N85, and carry on to the end of Alvington Road which finally emerges at the top of Carisbrooke. (The 'Waverley' is just down the hill).

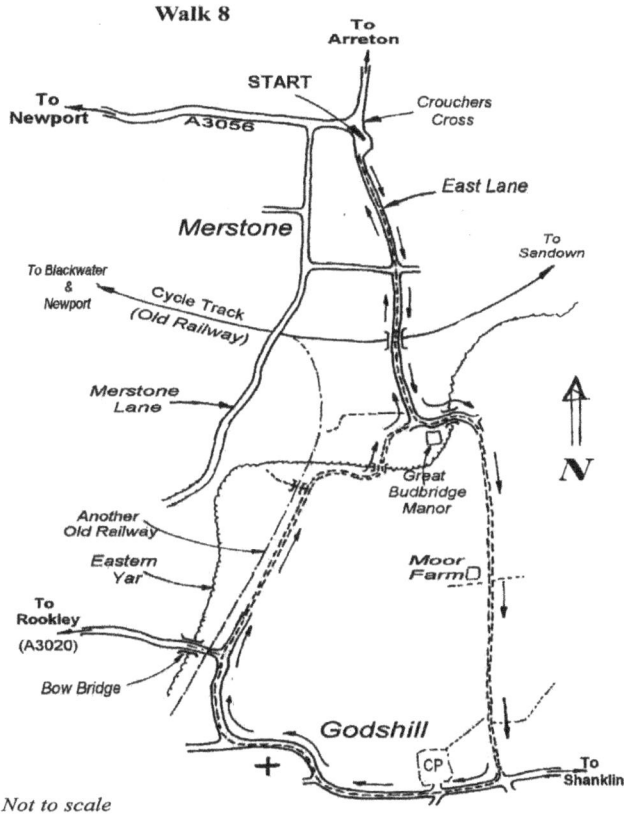

# 8. MERSTONE STATION TO GODSHILL STATION VIA PLUTO

*(A brief sojourn in 'Grockleville' and back again)    ( 4.9 miles)*

This walk starts from the old railway station at Merstone, near the split between the Godshill and Ventnor West lines. The starting point is now a car park, on what was the main line & signal box; the old island platform remains and behind is it a dip marking where more rails ran. Now the line is replaced by the Sandown – Newport cycleway, a popular 8 mile walk or ride. Head east out of here crossing the road, onto the cycle path N23 / footpath A4, and into Newlands. A few dozen yards along here the path splits; take the left footpath, with its wooden sign. The cycle track runs, with a few minor deviations, on the original rail route all the way from Newport to Sandown; it was opened in 2003.

The next significant point on your route, 500 feet on, is a bridge; a fine brick arch which carries the lane over, what was, the Newport to Sandown Railway, which ceased operation in 1956. The permanent way had been sold off to local landowners who had filled in the cutting and reclaimed the original fields. Fortunately, although taking many years of negotiation, the council bought back the rail route and turned it into a very useful cycle track; even re-excavating the cutting to do so and forming a ramp to give access to the lane.

Go up the steps here and turn right; the bridge is worth a pause, it's another good viewpoint; the land corner to the left beyond the bend in the cycle track, drops away to the marshes of the Eastern Yar, and the view to the right is up the wide valley to Newport.

After leaving the bridge, and passing the few houses to the left, you come to the delightful 'Little Budbridge Farm', some of the old barns are now in ruins but nonetheless picturesque, perhaps more so, as a result.

*The leaning barn about to fall,*

*Resounds no more on husking eves;*

*No cattle low in yard or stall,*

*No thresher beats his sheaves.*

From 'The Homestead' by John Greenleaf Whittier.

(Husking is the manual removal of the husks around the maize (sweetcorn))

After Little Budbridge, on the same side, there is an entrance track to some fishing lakes followed by a modern smallholding with a bungalow,

Budbridge Lodge. Some two hundred yards further on you will arrive at a guidepost at the end of a line of conifers on the right, which are acting as a windbreak for the adjacent greenhouses. Take the A22, signposted to Godshill, bearing to the left (here you join the 'Stenbury Trail'). The group of buildings to the right were part of Great Budbridge Manor Farm but are now, with added greenhouses and technology, part of the area's intensive market garden activity.

As you turn left you pass the old walled gardens of the Manor and a few steps further on Great Budbridge Manor itself appears. It is fronted by its wonderful pond, presently surrounded by a low metal fence. This surely must be one of the Island's outstanding settings; on one of my recent walks a lone shag flew over, no doubt eyeing the pond for fish; things must have been rough at sea!

Budbridge is another very ancient site; evidence of prehistoric habitation has been found here. Strangely it is not listed as a Domesday manor; my personal view is that this is because it would have been known at that time as East Merstone (or Merestone) and was given the Budbridge name later. No doubt part of the name refers to a bridge over the Eastern Yar here. There are a variety of theories with regard to the other part of the name.

The present house c1633 is usually referred to as Jacobean, which is not strictly true as James the 1st (Jacobus) died in 1625, therefore the house was built in the reign of Charles I. Just to confuse things the porch of 1688 could be called Jacobean (James II). I note, with appreciation, that the main gate has recently been renewed; a good design, well in keeping with the manor.

You next cross the Eastern Yar; the bridge is not in keeping with the area – very utilitarian! To the right of the bridge is a Water Authority gauging station which is certainly not in sympathy with the surrounding countryside. Also its rather derelict look doesn't help. (The Eastern Yar rises at Niton near Niton Manor Farm and runs through the Island's south east; it originally emptied into what was the greater Brading Harbour, but it now enters, much reduced, the harbour at St. Helens.) The track in the vicinity of the bridge has been raised and resurfaced, a considerable improvement; it used to be very muddy.

After the bridge an eastern aspect begins to open up ahead; across the fields a line of conifers act as another massive windbreak to row upon row of, from here unseen, greenhouses. Ignore the bridleway to the left (A 22 to Hale Common) and continue on the main track (now A49) which goes around to the right past a small lonely oak tree (ignore also the unsigned tracks to the left). The track now reverts to the natural sandy gravel surface and becomes quite open and exposed. A view gradually emerges away to the right front of Godshill church on its little hill. There is now quite a long section through rather bland fields, gradually rising although it appears flat as you go.

**At some point just beyond the highest point of this section of the track, but before the next hedgerow on the right PLUTO crossed this path.** Unfortunately, due to intensive farming, possibly including the removal of some hedges which may have been there in 1944. No standing marker can be found. I did find part of one lying nearby but it appears to have vanished sometime around 2003. Indeed, I have not found any markers for nearly a mile either side of this point; see also the return to Merstone Junction later in this walk and walk No. 13.

After passing the right hand hedgerow you arrive at a disused cattle grid, another good point from which 'to stand and stare'. The view to the left looks across to Waightshale and Lessland which were both old farmstead sites. Lessland farmhouse stands out solidly despite its exposed position. The position of Waightshale further to the left is now very difficult to locate from here since the last barn was demolished. In the distant view ahead it is possible to pick out the line of beech trees hanging to the side of Stenbury Down which mark the route of the old carriage road up to the Freemantle Gate.

Your track now continues, to the left of a hedge, down to Moor Farm, previously concentrating on the production of ham and bacon – but now producing acres of Fennel. Moor Farm itself is also an old site. The original 'hall' house had been converted into three cottages, but the present house of 1971 is built on its predecessor's position. Another footpath (GL 27) crosses your path at the farm, but you should continue south.

Having moved via FP A49 from Arreton parish into Godshill your path is now designated GL 46. Ahead the bulk of Stenbury Down becomes more obvious and the stump of the Worsley Monument can now be picked out on the highest point; it has been repaired after lightning, in 1831, modified the original. The path now descends off 'the moor'; turn right onto GL28 when you reach it, below the farm. It is a small path, signposted by only a small yellow arrow behind you.

(This avoids the main road, where the 'Stenbury Trail' crosses and as the GL 44 goes up through Godshill Park, past the beech lined hanging to the Freemantle Gate and on again to Appledurcombe or Wroxall. This was the Worsley's carriage road between their mansion at Appledurcombe and Godshill where they attended church.)

Your route, however, now goes to the right along the path a short distance to Godshill passing through the Old Smithy car park, with its coffee shop / restaurant and some other shops. If you choose to head down to the street, amongst the grockle shops there are two pubs and some convenient public toilets. There are also (unusual for these walks) plenty of places to eat. Beware; if it's a café you want; they close around four! There is plentiful free parking here, should you wish to approach this walk in a different order!

The return walk to Merstone is via Munsley Lane. There are many walks available from Godshill but to continue the PLUTO theme and return toward Merstone Station, it is necessary to walk through the village, taking the Newport (or Rookley) road. The road turns sharply right, however continue ahead to West Street (signposted to Niton/Blackgang, and proceed up it to Scotland Corner. There is a bend to the left, a footpath right, and another straight ahead, leading to Scotland Farm; this is the old station for Godshill. We will return to this point, but for now your route continues to the farm; head right when you reach it, and then left after a short while, and you will find the old platform to the left, within the greenery. Once you have seen the station house & platform, turn around and return to Scotland Corner, and take the path (now) to the left - GL23.

This leads back out to the north end of Godshill; proceed left along the low grassy bank by the main road, crossing at the bend ahead. Our path continues by taking the track which leaves the road to the right, from the apex of the bend, heading north; this is Munsley Lane (the A 2).

If interested, it is worth a short detour here for a few yards along the Newport road to Bow Bridge over the Eastern Yar (why 'Bow' I wonder?); but be very careful as there is no pavement along the road or refuge on the bridge. Let into the masonry of the parapet on the southern side is a unique milestone. The inscription was visible the last time I looked and read: - 'Erected in The Year 1769. To Newport Quay Five Miles'

Not by way of the river; that doesn't go to Newport! The Worsley's are thought to have been responsible for this and, of course, the bridge. The stone is also partly obscured due to the build-up of the road surface over many years. In full it used to continue with the mileage back to Appledurcombe!

On retracing your steps to the bend and the start of Munsley Lane it is to be noted that, had you walked this same short stretch some sixty years ago you would have gone under a railway bridge which took the Merstone to Ventnor West Railway over the road. This railway opened in 1897 and closed in 1952; a very short life!

However, returning to the bend, take the lane running north; it runs parallel to the old railway embankment on the left, which has been removed in some places but further on it has been converted into a private 'all weather' access drive to Little Kennerley. The track is very sandy and quite straight. On the right we pass Godshill's sewage farm and further over Munsley bog; next comes Munsley Farm. The terrain to the left is hidden from view by the old rail embankment and the high banks of the track; the track is used as a drive for a modern house which appears later on the right. Note the concrete fence posts along this bank, which are the old Southern Railway fence posts.

After this the track narrows and in a short distance turns left between the brick buttresses of a part dismantled rail bridge; before the owners made use of the old rail track this was the way into Little Kennerley. This is also a footpath (GL 24A) leading to Bohemia Corner; however, your route does not take this path but goes straight on past the old bridge on A 48. **It is in the vicinity of this old bridge that PLUTO crossed the path!** This is another location where I can find no trace of a marker; as there is a pond behind the embankment, and the railway had to be crossed, is seems obvious that PLUTO would have come beneath this bridge – no marker needed.

Your path, now much narrower, takes a sharp dive here, down to the level of the Eastern Yar which emerges from the left under an old rail bridge that remains intact, and is just visible through the trees. The path then turns to the right and follows the river, wandering along its banks. Godshill church, on its knoll, becomes visible across the marshland to the right. The path soon crosses the river by way of a relatively new foot/horse bridge. It then proceeds up toward the rear of Great Budbridge Manor, with glimpses of the architecturally interesting rear elevation of the manor.

The path now goes through the newer market garden which are part of the grounds; the path used to go closer to the manor, where there was much more of interest to see. (I wonder if it was formally diverted?)

At the end of the market garden area you rejoin your outward route from Merstone Junction to Godshill. At the guide post, on the corner of the conifer windbreak, another path to the left (A 22) could be taken to the Merstone road near Bohemia Corner, but the road from there to Merstone is too hazardous for walking. Continuing north on your previous route, Budbridge lane and then East Lane is, obviously, just a repeat of the previous walk in reverse except that the scene changes.

The high grounds to the north is usually the Island's central chalk downs; the exception is the down to the left of this view, St. Georges Down, which has quite a deep cap of plateau gravel lying on top of the chalk. This has been worked for many years; as can be seen from the machinery atop the down toward Newport. However, after the dip which accommodates the Arreton road the chalk spine continues to the east, broken only once more, by the Eastern Yar at Brading, before terminating in Culver cliff. On reaching the rail bridge, turn left before crossing it; this leads to a raised footpath parallel to the one we started on, or you might proceed right down to the cycle path. Either way, turn left/west to head back to Merstone Junction.

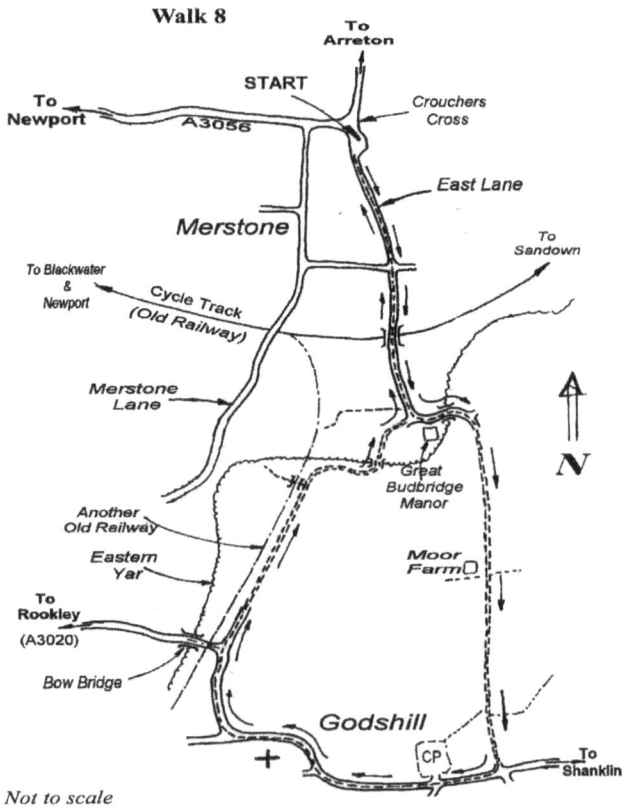

# 9. A NEW WOOD AND HUNGERBERRY

*(Another short walk in the Shanklin area)    (1½ miles)*

When I first set out to establish which Island paths had been crossed by PLUTO I was not to know that new paths would be created. However, thanks to the public spirited action of Mrs. A. Springman, owner of the Shanklin estates, some new woods have been planted which include some 'permissive' paths. Access to these areas has also been negotiated and this has allowed one more path to be included in the format of these walks.

The best place to start this walk is at Shanklin's old church of St. Blasius which has already been mentioned, as a minor deviation, during walk No.2. Walk straight through the churchyard, noting the impressive lych gate and the lovely little bird-bath on the grave to the right. There are probably many other things that will attract your attention.

At the back, or top, of the churchyard you will find some stone steps over the wall, quite rustic and often slippery. The signpost tells you that it is footpath 'SS10' to 'Wroxall and Ventnor over the downs'. (This is part of The Worsley Trail). The path takes you up, via a gate, through two fields. The first has been pasture in recent years and can be very marshy but either could be growing crops. However, the path is well defined and well used and there is a gate replacing a stile between the fields.

The path rises slowly with the wonderful sweep of the partly wooded downs above Luccombe on the left merging into Shanklin Down ahead with no apparent break; the road to Ventnor going up through Cowlease past Greatwood Copse unseen though probably 'heard'! Holme Farm is tucked away down to your left front. Cedar trees can be seen down by the road in front of the downs.

At the end of the second field a gate takes you into the edge of a small copse with some old ash trees. After a short rise you meet a more serious slope with steps. These steps take you up to an area of low scrub on a further though gentler slope at the top of which, in only a few yards, is a crossroad of paths. The older SS10 (The Worsley Trail) continues ahead, eventually even more steeply, but the path you need to take is that to the right.

At the time of writing there is a board here with an explanation of the new paths and the land which Mrs. Springman has allowed to be made use of for environmental improvement. Hopefully the vandals will allow the board to remain unsullied; with luck it is outside of their walking range! This has been the case for over a decade.

As you turn right you have an exceptional view of Shanklin and Sandown, backed by Culver Cliff with Bembridge and Brading Downs to the left.

The new path dives steeply down the grassy slope to the right. Beware, it can be slippery. As you go down it is necessary to bear to the left, skirting the bushes, aiming for the ash trees on your left. Here the way is slightly upwards, and over the boundary bank. It was difficult to comprehend in 2006 this was very ordinary open pasture, It has now been landscaped and planted with a variety of native trees criss-crossed by permissive paths, which by 2019 is looking quite mature.

Keep to the left-hand path which rises up through the wood a little way out from its border with the inland cliff which now becomes an impressive feature on your left. It never fails to amaze how those very large trees cling to the top of that cliff and how those at the bottom achieve incredible heights in their attempt to reach the sunlight from which they are sheltered on the south side. At the top corner you reach the highest point of the path, which used to have further grand views; perhaps from the conveniently placed bench. This view has been obliterated by the trees and has just become a nice spot in the wood.

From this high point turn down to the right, following the path around the border of the new planting; it goes downward very steeply here. In a while you arrive near the south-west corner of Hungerberry Copse.

The path now bears right, undulating down, with the side of Hungerberry Copse on the left. It was in this conveniently located copse that the huge 620,000-gallon header tank for PLUTO was constructed. This tank received all the fuel pumped across the Island from Thorness Bay and then fed it down to the pumps on Shanklin and Sandown fronts.

In this area it is noticeable how quickly nature is asserting itself with, despite the new planting, lots of self-seeded plants that include some brambles established amongst the younger trees. The primeval 'mare's tails', not unusual on the Island, are growing unabated. The path continues down the side of the copse. Really more of a wood than a copse, it seems to have been hardly touched since the removal of the tank just after the war. It contains a huge variety of trees and saplings.

When you reach the far corner of the newly planted area the path goes through the hedge where it meets the copse by way of two little footbridges over the ditches.

Over the bridge the way becomes a permissive path around the edge of the field, with the copse still to the left. **It is between the footbridge and the steps down into the next field that PLUTO crossed the path.** There is part of a marker in the margin of the copse that can sometimes be seen from the path; it depends on the state of the surrounding vegetation. The few wooden steps, which take the path down into the next field, go through the overgrown end of an old hedgerow, most of which has been removed to make one larger field. The path now skirts the east end of the copse until it runs alongside the garden of a house and then the inside of the hedge by the road where a, quite elaborate, combination of steps and stile leads onto Westhill Road.

Turn right in the road and go about one hundred and fifty yards up the road to the first turning on the right which is footpath SS85 leading up to Shanklin cricket club and the back of Shanklin Manor. As this path was part of walk No 2 I'll not dwell on it, **suffice to say that as you approach the end of the bushes to left and right PLUTO again crossed the path; there is a leaning marker in the hedge.** Follow the path around the back of the manor and you will arrive opposite the duck pond in front of St. Blasius' church and your starting point.

**Walk 9**

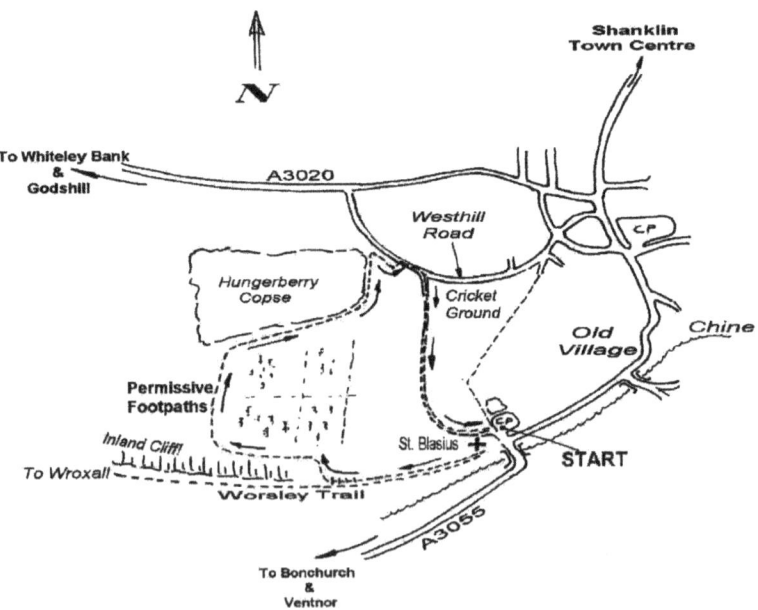

*Not to scale*

# 10. A FOREST CORNER FROM TUCKER'S GATE

*('New' fields and wartime dispersal (Just under 2 miles)*

Tucker's Gate is an entrance to Parkhurst Forest opposite Poleclose Lane on the Forest Road. The status of parking permission here is dubious. There is a bus stop nearby or one could walk from the car park in the forest – this is presented as an optional extension.

Why 'Tucker's' gate? I can find no answer to this. Parkhurst Forest was surveyed in 1793 by the head of The Ordinance Survey, Lt. Col. Mudge. Was Tucker one of Mudge's surveyors or was he perhaps, just the gate keeper? Or could it be named after James Tucker, the workhouse lad who later became a teacher. People from the workhouse were known to have worked on the road in the early 1800s. Presumably, at some stage, this was one of the 52 gates, which, it is said, had to be opened and shut during the journey on the carriage road between Newport and Yarmouth.

If taking the route from Parkhurst, park in the large car park on the right approaching Gunville Road from St Mary's. You then have quite a choice of paths; through the woods may be muddy, and may have squirrels, or the path parallel to Forest Road is a safe option. At Tucker's Gate, tucked (!) into the left hand corner near the road, is the start of your footpath. This is the N160, signed to Whitehouse Road This corner of the forest is mainly conifers but the path runs between an avenue of mature oaks, no doubt predating the planting of the conifers; it has boundary banks and ditches and is little used. You will notice that the path has, in the past, been wider than would be needed for just a footpath. This is because Tucker's gate was a fork in the 'main road' and this path was, in the 18c, the road to Newtown. It joined up with Colman's Lane when Whitehouse Road was just a track leading north, on the edge of the forest, from Vittlefields Farm.The time scales are a little different but Kipling had a similar situation in mind when he wrote: -

> 'They shut the road through the woods
> Seventy years ago.
> Weather and rain have undone it again,
> And now you would never know
> There was once a road through the woods
> Before they planted the trees,
> It is underneath the coppice and heath,
> And the thin anemones.'
>
> From 'The Way Through the Woods'

The path first skirts the back of the Island's waste treatment plant, which has recently been 'improved'. You then come to what used to be another major gate into the forest, as will be seen by the approach drive from the road which comes straight through the industrial site on the left, to the unused gates in the boundary fence. The way into the forest from the gates is now merely a narrow path. The site is presently in use by the builders' merchants Sydenhams but was built after the bombing of Cowes in 1942, as a dispersed part of the Saunders-Roe aircraft factories. No doubt the site was chosen for its rural position and concealment possibilities. After the war, until 1965, it was a major Saunders-Roe machine shop; always referred to as 'Forest Site'. (Sometimes also known as 'Forest Side').

Your path goes straight ahead now, thankfully less noisy and more like the forest that you would expect. On my last visit here I was escorted by a barn owl flying up and down, despite it being mid-morning. I was obviously seen as an intruder.

There has been some logging done in this corner of the forest, rather untidily I thought, so it thins out a little as you come to the end of the trees. At the edge of the forest, a mini foot bridge and a stile take you into open fields.

Just before the stile is a pair of quite elaborate wrought iron straining posts which once formed a gateway for the path and form evidence that this has been a footpath for very many years. It is possible that these posts actually date from 1815, when this became the new boundary after they reduced the area of the forest by over 1000 acres.

The fields that the path now goes through are relatively new in that they date from that reduction of the forest in 1815. It is difficult to imagine the effort that went into clearing them using only horses, chains, and manual effort! The path goes at approximately 45 degrees to the right across the field, following the line just taken through the forest, and as you reach the crest of the field you will be able to see the gateway to aim for, a little way back from the far corner. **It is between the centre of the field and that gateway that PLUTO crossed the path.** I have reason to believe that it crossed nearer to that gateway ahead. There being no hedges either side of the path there would have been no need for markers.

At the gate there is a stile on the right-hand side which takes you onto a virtually unused track which was once the access to Forest farm, on the right, from the main Newport to Yarmouth road.

Your route now is across the track to a gate opposite; at the time of writing there is no stile. At this point the path leaves the line of the old road that went diagonally across the present field to the right. The path now crosses two fields, hugging the field boundaries to the left. Initially, in the first field, there is a short length of track before entering the field proper. The frontage of Forest Farm can now be seen on the right.

This is flat country but to the left are the Downs; Alvington, Bowcombe, Rowridge with its TV mast, and Brighstone, with its own forest. A ditch and another stile separate the two fields. In this second field the view to the right opens up with Sand Hill near to the forest and the quite grand looking farmhouse of Sandhills Farm.

At the end of the field a stile takes you into Whitehouse Road, the well-known 'Rat Run' for those trying to miss the congestion of the Newport to Cowes road. Having arrived at this seemingly obscure point it is difficult to understand why the path has survived, when the old road fell into disuse. I suspect the reason is that, in the fields across the road from the stile, by a tiny stream, were Colman's Cottages, now gone. Why via Tucker's Gate? Probably because it's opposite Poleclose lane, which would be the best way to walk from Gunville and Alvington Manor.

Turn left at the road, but from now on take more notice of the traffic than anything else. As you walk, hopefully on the verge, you will notice very shortly a large surfaced area in the fields on the right complete with stillage banks. This is the site of one of the Island's exploratory oil drillings. They were rather guarded as to what was found but the fact that it hasn't been returned to nature suggests that there may be something useful down there when, or if, it becomes economical to extract it!

This long straight road is rather featureless but there are some points of interest. The road, when a track, used to veer away to the right here and join the main road at Vittlefields Farm it was probably given its present alignment in 1815. The land either side of this piece of road has been given over to horses or 'set aside' like a great deal of the Island. The area up ahead as you approach the main Newport to Yarmouth road is known as Vittlefields Cross. The Farm is further along the main road to the right. You first come to a house on the left. There used to be a pronounced ditch alongside the hedge by the house and it was quite normal to see a few bikes left, quite safely, in this ditch where people from Porchfield and Newtown had left them in order to catch a bus into Newport. (They may still do so.)

This was of course before most people had cars and for many years now there have been some bus services to these parts.

The name Vittlefields seems to be a bit of a mystery for the experts. The 'field' bit suggests that this was a clearing or the edge of the forest, and the 'V' in front was an 'F' from its first recording in around 1300 until about 1600 when the 'V' appeared; probably just the result of an error of speech. One theory concerns fiddling of the musical variety but the festivities, to do with the forest and greenery, cited as the possible reason, were usually done much closer to the town. Over to you.

The road continuing straight ahead at the crossroads is known as Betty Haunt Lane, a name derived from much corrupted origins. You need to turn left here onto the verge of the main road, but before doing so look across diagonally to the corner of land to the right of that opposite lane. There is a large area of hard standing, which was visible for many years but has now been sold off and partially fenced in. This was the site, from 1943 to 1945, of two Type 'B' Robins hangers known as 'Vittlefields 1 & 2'. These were not used to house complete aircraft but were additional dispersed factory space for Saunders–Roe Ltd. to maintain wartime production. They would have been provided by The Ministry of Aircraft Production (MAP) and made by their contractors, Dawnays.

You are now faced with a somewhat tedious plod up the verge of the main road; this part of the walk can be a bit of a 'trial by traffic' so beware, the traffic is fast and furious here. So why do it? Well, it's really in an effort to be as complete as possible; and, as you will notice, it is particularly significant with respect to the alignment of the pipeline.

It is unlikely that this is the route of the road which the governor, Sir Thomas Holmes, had made through the forest in 1670 to facilitate King Charles II progress from Gurnard to Yarmouth, as he is said to have gone via Great Park. It just doesn't fit. However, with the passage of time and accuracy of recording, it is possible that part of it may be.

The road, although old in its general alignment, was almost certainly given its 'new' straightness following Mudge's Survey of 1793. Much of the manual labour in building the road is said to have been provided by inmates of the Island's workhouse at Parkhurst. (Still standing as part of St. Mary's Hospital) The fields to the left are newer still, having been formed after the 1812/15 changes to the forest boundaries. The fields to the right were cut out from that part of the forest bordering 'New Park' a little earlier, in the late 18th century.

When you reach the top of the rise you come to the entrance to Westwood Farm, on the left, this is the Forest Road end of the track that you crossed earlier in the fields. Immediately after this is a new entry for something not yet apparent. **It is about halfway between here and the houses on the left a little further on, that PLUTO crossed the path (In this case road).**

There are good markers on both sides of the road. This is something of a surprise because Forest Road, being on poor ground, has had a huge amount of work done on it since the war. However, despite this it obviously hasn't been widened to any great degree and the hedges must be pre-war.

The lane a little further on to the right, not on this walk, is the access to Green Park and Cooks Farm. (This is a footpath N70). Next to this lane are two establishments which have been developed on the site of two more of those Type 'B' Robins hangers known as 'Green Park 1 and 2'. They were placed there for the same purpose as the ones mentioned earlier. The second of the two existing buildings, presently a BMW showroom, is in fact one of those hangers; it retains its original shape but has been added to and reclad more than once. Opposite this site the forest starts again and becomes the left hand boundary to the road.

Continuing along the road the old Saunders–Roe factory in the woods, now Sydenham's, soon comes up on the left. In 2006 the entrance still has the wartime look about it with its guardroom by the gate. The brickwork on the front looks newer than some further back. Next along the road on the same side comes the aforementioned waste disposal plant and then it is only a few yards to your starting point at Tucker's Gate.

# Walk 10

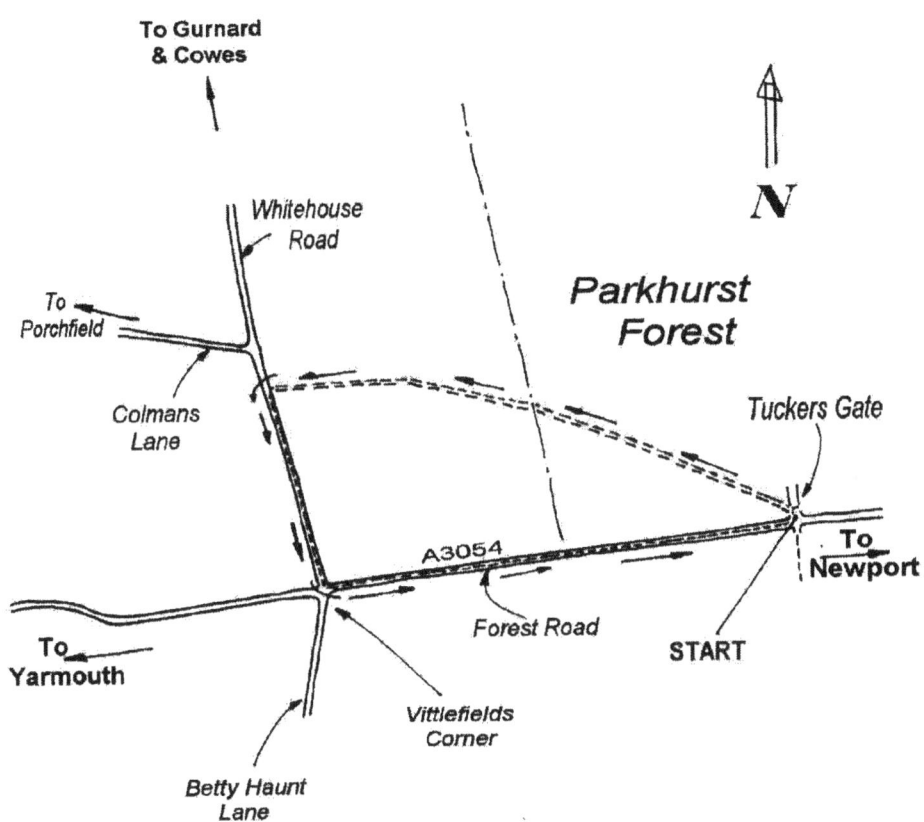

*Not to scale*

# 11. TO RYDE FORD AND BEYOND.

*(A Circular Walk from Godshill)*   *(Nearly 4 miles.)*

An apparent misnomer this; but all will be revealed later. A convenient starting place for this walk is the main car park at Godshill, but be aware that it closes at 6pm.

Toward the end of the car park on the right hand side looking from the entrance, a stile gives access to footpath GL 28 which leads up across a field, often occupied by pigs of the Island Bacon Company; never fear, they are fenced in.

The top of the field, surprisingly, gives a view of a considerable amount of the Island. Back to the left is Godshill church on its little hill and from there in a broad sweep can be seen the two TV masts on Chillerton and Rowridge Downs followed by the Downs of St. George, Arreton, Mersley, Ashey, and Brading straight ahead. Slightly to the right in the middle distance the red brick of the old farmstead of Lessland is prominent; rather taller than most farmhouses of the same period. Behind it is the higher ground around Newchurch and away to the right Shanklin and Wroxall Downs followed to the right rear by Stenbury and Gatcliff.

Continue across the field, now downhill, to Moor Farm lane. Across the lane to the right is a stile into a pasture. Keep to the boundary on the left, the path, then takes you around a left hand corner to a gate after which the line of march is diagonally across a very wet pasture to the fence on the right hand side which runs alongside a stream. This chalybeate stream rises in Godshill Park and runs over iron bound sands as can be seen from its orange colouring. There is a stile with a very rickety bridge in the right corner – DO NOT try to cross here! Keep the stream on your right. The areas of reed beds are a wonderful haven for birdlife and the fields due to their dampness are often covered with rooks and seagulls. As a further stile is reached, now on the right hand boundary you may wish that you had wellington boots!

This stile leads into further pasture, with another crossing to ignore over the same stream and, near the end of the field, the path crosses a farm track coming down from Moor Farm on the left. The track continues across a vehicular bridge to the right, but not the footpath. The footpath continues ahead over a further stile and eventually turns right out of the field over a stile and footbridge over the same stream.

Since crossing the farm track the path has been in a narrow strip of land once known as Ryde Mede (or Tydelingeham) which stretched from here to the Eastern Yar; it was owned by Quarr Abbey.

Somewhere in this vicinity, either by the vehicular bridge or between it and the footbridge, was a ford over the stream once known as Ryde Ford. Nothing to do with the town of Ryde, both places attracting the same name due to their being to do with a small stream.

After crossing the footbridge and passing through a thicket of blackthorn the path now joins that farm track at right angles. A turn to the left onto this track leads you through a completely uncultivated area of scrub and marsh rich in plants and wildlife. This continues for a few hundred yards with a field drain to the right. The stream that had been crossed by the footbridge is now out of sight as it veers away to the left across the marsh, eventually joining the Eastern Yar some half a mile away.

At the end of this area the footpath stays on the same track which now becomes more formalised and proceeds between banked hedges bordering cultivated land. The path does not swing around to the left on the grassy track which it may appear to. **It is at about this point, where the path enters the hedged in track, that PLUTO crosses the path.** I have been unable to find a marker here and it becomes part of the longest gap between markers that has been encountered (Between Bohemia in the west and Summersbury in the east).

You are now in a lane, leading to Lessland, it is obviously of ancient foundation and would, no doubt, have been, via Ryde Ford, the route for the residents of Lessland to church at Godshill. Gaps in the hedge to the left and the more open aspect to the right give some views over the surrounding countryside. To the left front, glimpses of the vast glasshouses of the Arreton valley can be seen together, at times, with fields covered in plastic; one imagines that one day there may be more plastic remains than earth!

Lessland farmhouse now appears to the right front. This was Liscelande or Litesland in 1086 and has two entries in the Domesday Book, one saying that it was held by the King the other saying that it was held by another William called FitzAzor. Perhaps he held it for the King! It has always been an important arable holding. The present house is said to have been built in 1722 with major modifications to the front in 1829. Regrettably the associated barns, some of which predate the present house, do not do the house justice, having been allowed to fall into a ruinous state. It is probable that the house is no longer connected with the farming activity.

The path now skirts the buildings to the left and continues with a rather disjointed hedgerow boundary on the right. About two hundred yards further on the path follows the track through a major gap to the right and crosses a strip of land to the corner of an agricultural reservoir; it then continues up its left hand side. This fairly recent reservoir now seems a mature part of the landscape and has become quite a wildfowl habitat; even occasionally some waders. There is a 'top path' which yields nice views of the reservoir.

On the left is a well-established hedge and ditch but at the end of this field and some way past the end of the reservoir the track jinks to the left and the field boundaries to two or three paddocks appear on the right. Go past the paddocks; the track emerges onto Bathingbourne Lane. Bathingbourne also appears in the Domesday Book but the farmstead associated with it is along the lane to the left.

A right turn into the lane takes you due south. This is normally a quiet country lane but beware the odd trade van taking a 'smart' short cut. Just before the first bend near the houses on the right is an unusual bridge, the watercourse beneath is often dry, but to the left there appears to be a masonry-bounded holding pond and to the right of the bridge and the remains of a sluice. One ponders the reasons for these works, which are obviously quite old; local knowledge suggests it was the old reservoir, used to flood the fields when needed.

Do not take the footpath which goes straight ahead but continue around the bend to the right. Also on the right is Green Acres Farm which is presently the home of some alpacas; llama type animals normally found in the Andean region of South America.

After this the lane rises through a cutting; make sure you walk where you can be seen as the lane is narrow and there are no verges in this section. On the other side of the hill to the left side of the lane is another establishment presently dealing with rare breeds, 'Badgers Farm'.

There used to be some buildings here of possibly military origin (they were of a type known as 'half brick huts'), but I was never able to establish their purpose and they have been rebuilt in recent years. Somewhere in this region, it is thought more to the north, there was reported to have been a wartime Radar installation; this just may have had something to do with it! There is a new build here now.

The lane wanders on and is shortly joined by the track from Bobberstone on the left and almost immediately, on the same side, a white painted house known as Summersbury.

**Between this house and the lane to Lessland, a little further, on the right PLUTO crossed the path (in this case the lane).** There is a good marker here on the left side of the lane. In theory there should be some markers some way down the track leading to Lessland; but none are to be found, and the owner doesn't know of any.

The lane continues up, down, and around but nominally due south. The typical rolling English road:

> *'Before the Romans came to Rye or out of Severn strode*
> *The rolling English drunkard made the rolling English road'*
>                                    G K Chesterton

How boring it would be if all our roads were Roman ones!

Just after the entrance to Froghill on the left a footpath, the GL31, crosses the lane. Take this path to the right, it is an obviously old, part sunken avenue which, after a short distance, dives down and turns abruptly left alongside a little stream at Sandford. This is another stream which has risen in the upper reaches of Godshill Park and used to power a mill which once stood in that park. Very noticeable here is the little footbridge which has looked as though it has been about to fall down for the last fifty years; a great tribute to the craftsman who formed its brick arch. It looks, even in its present state, so much better than a concrete pipe.

Sandford is quite small now in relation to its neighbour, Godshill, and is unusual in that it has retained its present name since its entry in the Domesday Book. No doubt the stream, where it crossed the main road ahead, was forded then; it also, at that time, powered two mills, so it may have been somewhat larger than it is now.

The path now goes behind Sandford House and comes out onto the main Godshill to Shanklin road. There are options here on how you complete the walk; you could follow the road to the right into Godshill but there is a very dodgy section just short of Godshill, called Poor House Shute, where there is no pavement and traffic is virtually continuous.

The best solution is to extend the walk by crossing the road and taking the ramp up to footpath GL42, to Godshill Park, which starts from the opposite bank. This path initially goes along the left-hand side of three or four horse paddocks before coming out into the more open fields of Godshill Park. At the end of the last paddock go through the bridle gate and keep to the hedgerow on the left. Further over to the left is the shallow valley which was the site of the water mill in the park. The hedge terminates at a gateway onto the drive of Godshill Park Farm.

Turn right on the drive which, in a few yards joins the old carriage drive through Godshill Park. This drive was established by the Worsleys of Appuldurcombe to connect their mansion with the village of Godshill. It is believed that the park was once part of a more extensive Appuldurcombe park. The Park House opposite the farm drive appears to be Victorian and almost certainly postdates the layout of the park. Turning right onto the carriage drive the way goes through the remnants of the once grand parkland which has only been fenced in fairly recent times; previously it had been left to the cattle grids to control the movement of stock.

Unfortunately, only one or two large isolated trees remain to remind one of classic English parkland. Looking to the left front, just before the descent down to the main road, Godshill church can be glimpsed, in wintertime, peeping over the hill.

As the grid is crossed, when arriving at the main road, note the large block of concrete forming a raised platform by the fence to the right. This is a milk churn stand, where farmers put their full churns of milk so that they could be more easily loaded onto the collecting lorry; Not required since the advent of bulk collection in the mid 1970s.

Turn left at the road, crossing to the pavement on the right hand side, and continue into Godshill. The next turn to the left leads into the car park, completing the circular walk.

# Walk 11

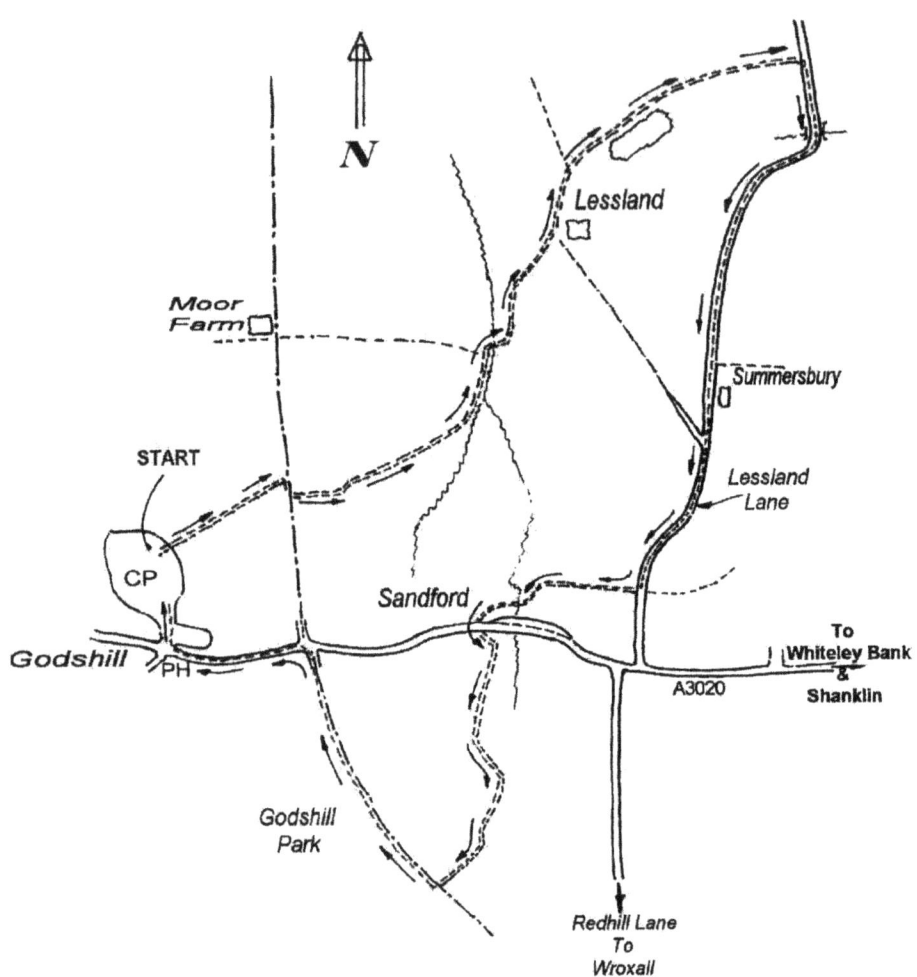

*Not to scale*

# 12. UP AND DOWN A DARK LANE

*(A walk south of the castle)*     *(About 3½ miles.)*

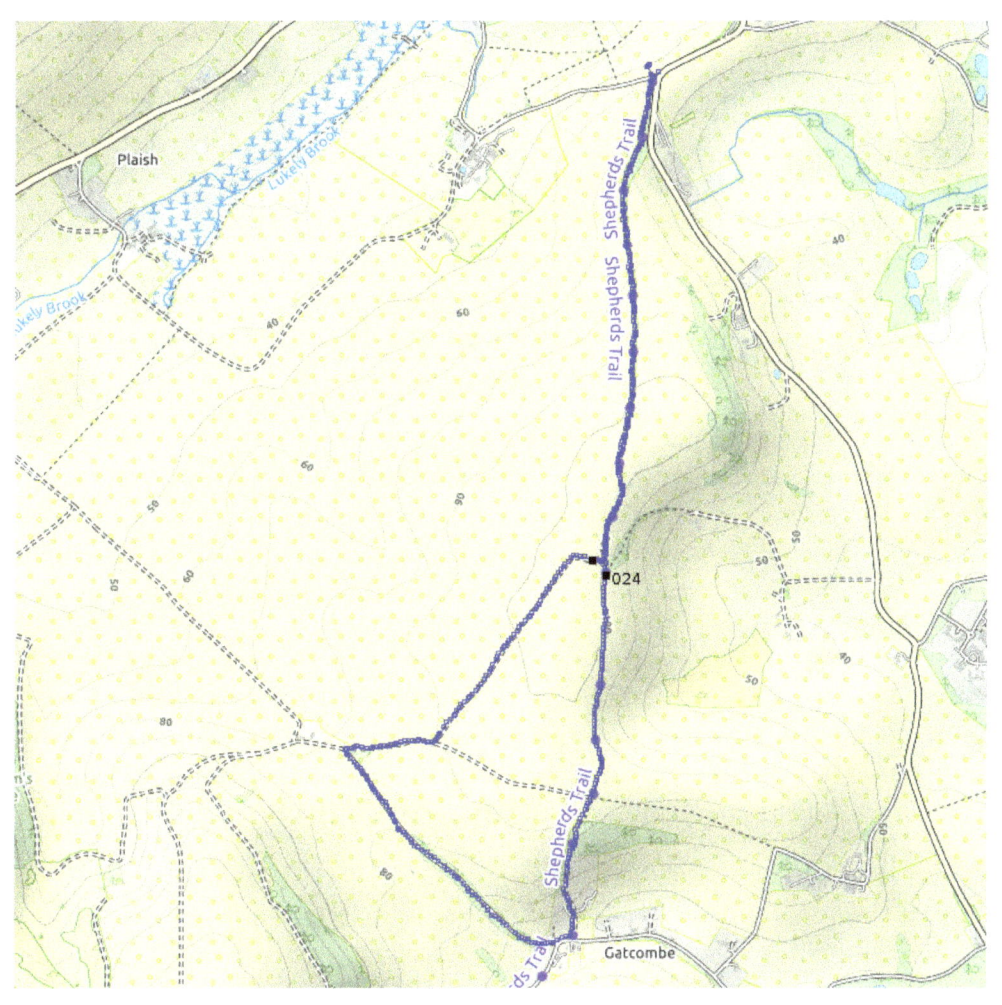

This path, one of my favourites, locally known as Dark Lane, is part of one of the Island's long distance trails designated by the County Council as 'The Shepherd's Trail'. It starts from Whitcombe Cross as footpath N108.

To get to Whitcombe Cross from Newport one can walk over Mountjoy from Whitepit Lane via N24/26; this will bring you to the Priory on Whitcombe Road where you turn left for a short distance along the road to Whitcombe Cross.

If starting from Carisbrooke, go via the castle or Clerken Lane. (N 66) (The Shrubbery) to get to the N88 leading south to Froglands Lane, then turn left to Whitcombe Cross. Alternatively, a car can be left at the Car Park (provided by the Rees Jefferies Road Fund) opposite the Priory on Whitcombe Road.

Initially Whitcombe Cross appears to be no more than the junction of the main road and the lane to Froglands but the path you want is the N108, between these two, to the south, off the field entrance. (Beware 'Perry's Guide' here; Editions 1 to 7 show a junction a little further to the south that doesn't exist).

To start with, the path has no great attraction, its line has had to be modified a few times over the last thirty or so years due to a tendency for it to slip into Whitcombe Road below, from which it ascends at a narrow angle. However, after a short distance it resumes its original track and form. It now shows itself as a classic sunken track, probably much older than the road it has just left.

Was it perhaps this track that Edward Worsley and Richard Osborne used to conceal their approach to the castle when bringing the horses in the March of 1648 during the failed attempt to spirit King Charles away from imprisonment? There is no doubt that there are feelings here of times past!

At the time of writing the path is in good condition, if anything a little too clinical, and perhaps belying its history, because in recent years it has been made suitable for bicycles. Previously the surface changed much more with the seasons, sometimes a veritable riverbed with considerable steps in the bedrock to negotiate, which were then evened out by soil eroded from banks and field together with leaf fall and twigs.

Another aspect which changes over the years is the degree of 'darkness'. I have known the hedgerows, high on the bank, cut right back but now it is at its most mysterious with branches of hazel, ash and oak meeting overhead to form a continuous tunnel. What a wonderful route it was for those, who in the past, would have wanted to move toward Newport unseen. No doubt used by smugglers, coming as it does from the south, heading for the grog shops of Newport: -

>'Five and twenty ponies trotting through the dark
>
>Brandy for the parson, baccy for the clerk'

From 'A Smugglers Song' by Rudyard Kipling.

Like many such sunken lanes, one is always in awe as to how the bushes and trees manage to hold on and indeed thrive, with so many of their roots exposed and so little to hang onto. Wet years produce a wonderful array of mosses and fungi on banks and roots, which must be of great delight to those interested in botany.

Unfortunately, rabbits seem to be the greatest threat to the banks, and to the whole character of these sunken ways; whilst their excavations are part of the character, they have become a serious problem and, if not controlled, will cause many banks and hedgerows to collapse. (Is it due to affluence that we are not eating enough rabbits?)

This path can be a delight whatever the weather; even when a storm is raging one can feel protected from the worst of the rain, wet maybe, but not buffeted. In sunshine it's just magical. On a recent visit I found only one relatively modern item to break my dreams of days gone by and that was a disintegrating roll of rusted 'Dannert' wire (as used in barbed wire entanglements) thrown up in the bank near the end of the 'tunnel'. That, I suppose, had its own nostalgia in that it was probably discarded when the Island was cleared of its obstructions to the enemy in 1945.

This sunken section of the path continues for about half a mile and when the sun is shining you can see the brightness at the end of the 'tunnel' where the path emerges into the high open fields. At this point I am always reminded of part of one of Winston Churchill's speeches during the dark days of 1940. The context is, of course, totally different but the words are so apt: -

>'----- through the dark valley we can see the sunlight on the uplands beyond'.
>
>WSC. July 14th 1940

As you break out into the open you have wonderful views to the east right round to the northwest. The height here is about 345ft (105m) so the view to the south is taken up by the bulk of the nearby downs variously named Garstons, Dukem, Newbarn, Westridge and Chillerton, all around the 540ft (164m) mark. To the east the depth of view is much greater; eventually you can see right out to the Island's south-eastern extremities from Culver Cliff to St. Bonifice Down.

The path now runs between the headlands of two fields until it joins a hedgerow on the left; the other side of which is a drop of some ten feet.

Just before the path dips into a trackway, a footpath is signposted to the right, this is N107/G23 towards Garstons; more of this path later. Follow the main path down into the dip where it crosses the head of a track coming up from Vayres Farm on the left. **As you start to rise again from this junction you pass the point where PLUTO crossed the path.** There is a marker here, nominally complete and in reasonable condition for a seventy-five year old; if you can find it!

Rising out of the dip the path continues through fields with the hedge on your right. There was once a gate at the top of the rise that marked the point where your original path, the N108, becomes the G6 as you will have just crossed the parish boundary from Newport into Gatcombe. In the next field the hedge becomes a shallow inland cliff, not unique on the Island but relatively unusual when bordering a footpath. It's quite eroded but again it's the rabbits causing the worst damage. I always have a close look to see what has been exposed, but have yet to find that hoard of gold coins.

As you rise out of this field and through another gate, take the opportunity to admire that view of the east of the Island. When underway again you will arrive, after a few yards, at a crossroad of paths where the G6 crosses the G10. West would take you to Garstons and Westridge Downs or through to Bowcombe on the Shorwell Road. East would take you via Hill Farm to the Whitecroft, Gatcombe road. On this occasion go straight across and continue south to the 'top' of Gatcombe.

The first section of this path goes very steeply down through the edge of a copse; please take great care here as it can be very treacherous underfoot. After a dwelling on the left, the path is partly concreted. (Sometimes helpful, sometimes not!). Gatcombe lies quietly in the valley to the left and Tolt Copse is prominent on the hillside opposite. The pair of houses high up on the right a little further down are 'Seely' cottages. The Seely family built cottages of this standard design, for their workers, on many of their Island properties. Opposite these are the Isle of Wight Hunt kennels. Another pair of cottages, lower down on the right, are also 'Seelys'. Note the thoughtful positioning of these two pairs of houses.

There is much to be said of Gatcombe but as this walk only clips a corner of it I'll be brief. Gatecome in the Domesday Book; the name said by the experts to refer to the valley where goats were kept rather than the more obvious gate to the valley! Certainly the contours here would support this. The church, rectory, and mansion are all closer to the main road to the east and worth a quite separate wander around.

At the bottom of the hill you emerge onto the road through Gatcombe with some, unfortunately rather modern, barns opposite; it is no longer the through road that it used to be when it linked the Chillerton road to Bowcombe on the Newport to Shorwell road. Most of Gatcombe, and there is not very much of it, is along the road to the left. Turn right here and keep right into Snowdrop Lane with the much-modified Snowdrop Cottage in a rather idyllic setting on the left. (That lane continuing to the left only goes to Newbarn Farm but there are paths branching off from there).

The lane you now enter lies in a deep cutting; a testament to its antiquity. The lane really only goes as far as Garston's these days but it was part of that old through road. There are some fine old trees here in this sheltered hollow. One old oak has recently been given a nesting box for owls. In snowdrop time this part of the lane really is a picture.

Eventually the banks decrease to some normality, the hedges are more of dogwood and hazel, and glimpses of Garstons Down appear to the left. There is a small group of trees on the right as you approach Garstons' Barn and at the end of the trees you need to take the bridleway which goes sharply back to the right. (This is G10).

As you turn, the entrance to Garstons is ahead and the metalled lane continues a little way to the side of Garstons itself. You may like to take the diversion of a few yards to the end of the lane which stops at the barn. A bridleway (G7) goes up over the down to the left. The old road used to go straight on but that has been obliterated in favour of another bridleway (N202) after that kink to the right. Although long known as Garstons Barn it was a farmstead well into the last century.

The place is almost certainly associated with John Garston an Island landowner who died in 1431. His properties were later absorbed into those of the Bowermans by marriage but from about the mid 17c the holding was farmed by the Urrys of Sheat and Hill Farm. Mapmakers of the past raise a slight question regarding its origins due to referring to it as 'Gunsons' but I suspect this is more likely to have been an error by non-local mapmakers; not uncommon.

I remember enquiring about it, when it was for sale, in the very early 1970s; the freehold was going for £12,000 but it had no mains services and water was from a well by wind pump. It was eventually sold and modernised, but now (2006) it is being extended and converted into a veritable little palace. Hasn't quite got the original charm though, that will take a few centuries, but it has a glorious view over the upper Bowcombe valley.

However, back to that turn to the right onto the G10 heading east. This is the bridleway you crossed earlier before going down into Gatcombe.

This was also a carriage road going west to Bowcombe but by-passing Gatcombe. Gaps in the hedgerows a little way up give some wonderful views; across to Bowcombe Down on the left and as far as Stenbury and St. Boniface to the right.

After some two hundred yards turn left into another sunken bridleway, this is G23; it is not signposted but it's the first on the left. This very pleasant path leads up past fields called Garden Field on the left and West Croft on the right to a field at the top which Bill Shepard tells us is called 'Gallants'. The way used to be quite rough going but in November of 2006 it was bulldozed to its original cart width, which has rather upset its character. I fear the banks are now more likely to collapse into the path. To me, not an improvement but no doubt it will feel more like its old self when nature resumes ownership.

A short distance up the path you cross back into Newport parish and the path becomes the N107. The hedges are of hazel, hawthorn and maple but eventually the left-hand hedge fizzles out, followed later by that on the right. The views are extensive; called 'Gallants' (There is another field with the same name near Gurnard) says Bill because from here you could see the top gallants of full rigged ships in the Solent, at Spithead and even in Sandown Bay to the east.

Eventually the well-worn path takes a turn to the right. (Maps will show the path going straight on here, cutting off the corner, but it's no problem to use the obviously worn path and go around). It's only a few yards. **About halfway between the right turn and the turn to the left at the field boundary ahead PLUTO would have again crossed the path**. Any hedges there may have been in this area have long since been grubbed out so no markers exist.

At the field boundary you will find yourself above that dip in your outward route where it was joined by the old cart track coming up from Vayres. You will need to turn left along the boundary for a few yards to join up with your outward path. There is a signpost marking he junction as previously mentioned. Now it's simply just a matter of retracing your steps down Dark Lane on 'The Shepherds Trail'; it is just as nice going down, there are places where fine views can be had of Carisbrooke Castle and you are bound to see something that you missed coming up. Take care at the bend in the road as you emerge at the bottom, before you gain the left-hand verge for your return to you starting place.

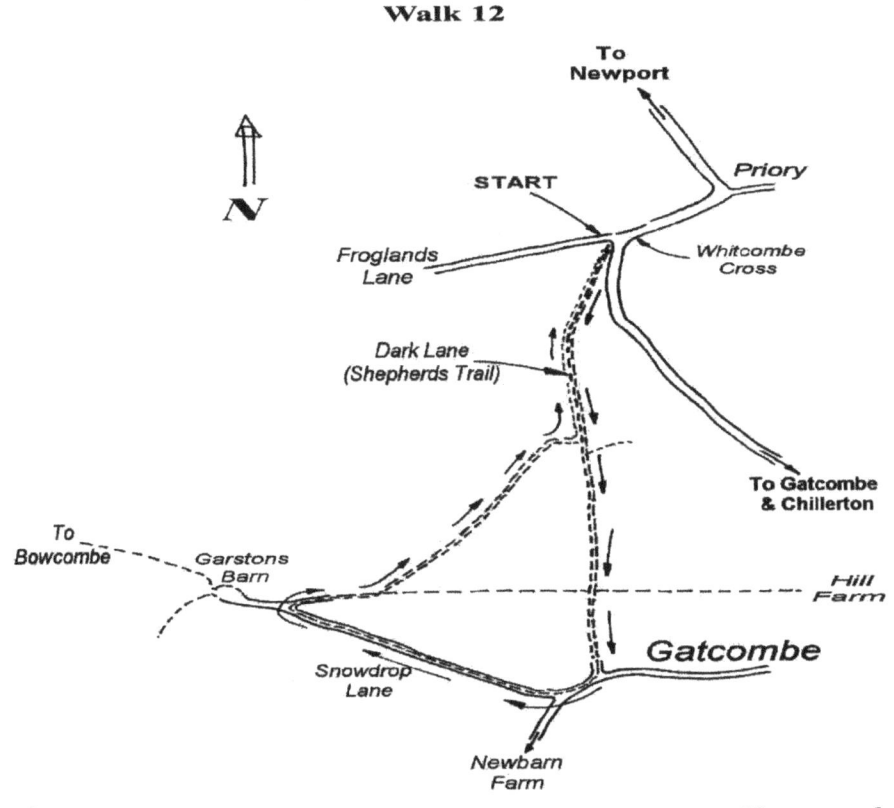

Walk 12

*Not to scale*

# 13. ALL AROUND WHITELEY BANK.

*(South eastern farms, downs and mills)*  *(Nearly 3½ mls.)*

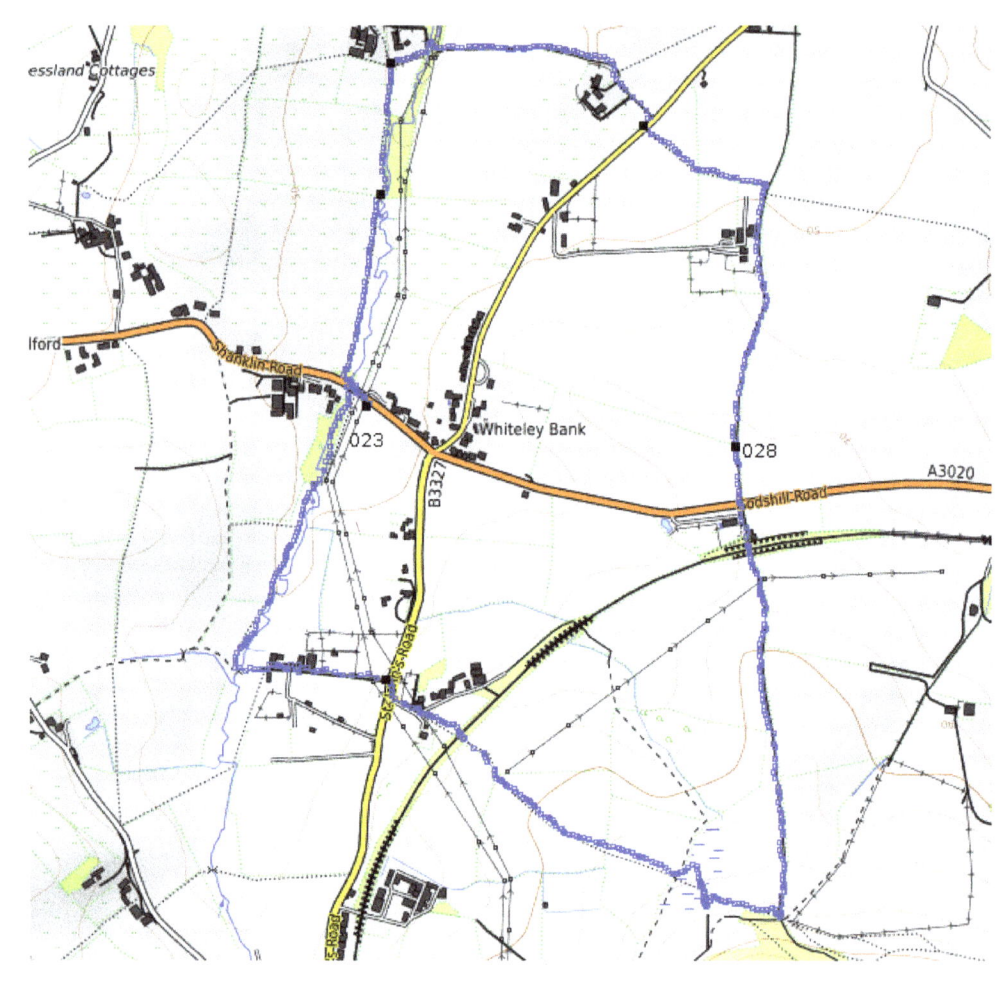

This walk starts just to the Godshill side of Whiteley Bank where there is a shallow lay-by on the southern side of the road. Cross the road to the north side, on the Godshill side of the bridge, and you will find a stile for the GL29 footpath to Bobberstone and Bathingbourne. If you have done the 'Criss-Crossing Sandford' walk you will have been across this field before, but a little to the left.

This path follows the stream along the eastern edge of the field. A rather unruly stream, it twists and turns, and nibbles away at the field and the sandy soil is easily eroded. Quite a lot of small willows line the banks.

At the end of the field a double stile, with the stream here some six feet below, takes you into the next field. **It is about here that PLUTO crossed the path**. A few years ago part of a marker could be seen from the second stile, lying in the banks of the stream; it may come to light again one winter's day when the nettles have died down. Did PLUTO go below the bed of the stream here? Or did it dive across between the banks? There is some old, mossy concrete down there!

Continue through the next field, hugging the right hand hedgerow. There appears to have been a track alongside the stream here on a lower level than the field and may have been the original track but it is mainly overgrown now. Judging by the trees on a definite bank, lining the stream, this was probably an old boundary and may still be.

At the end of the field near some mature ash trees the main stream takes a ninety-degree bend, away from you, to the right, along with the embanked boundary. The route however carries straight on over another stile and ditch and continues, now enclosed, with a mixed plantation of young trees forming a boundary with a lower meadow to the right.

Over the next stile (ignore the stile to your left that is GL31) is Bobberstone, it has had various spellings over the years but the present one is much the same as the one first recorded in 1459. Your path goes to the right, over a stile alongside the field gate and crosses the end of that lower meadow in front of Bobberstone farmhouse. A little modernised but still delightful in its setting; with its surrounding ditches it could almost claim to have a moat. Across the meadow in the left hand corner a further stile gives access to a short link path to the junction with the bridleway NC33, onto which you need to turn right across the substantial bridge of ex-railway sleepers. This is now the main stream again, which you have followed from the road.

Worth a pause here; the 'newness' of the bridge doesn't detract from the beauty of this little glade.

Underway again you climb away from the stream into open fields with, from the right, a view of Week, Wroxall, St. Martin's and Shanklin Downs; to the left Ashey Down with its white sea-mark. Conifers peep above the brow of the rising ground ahead. At the top keep to the left of the hedge which comes into view. This is another banked hedgerow, undoubtedly made up from past field clearing. ie. Stone picking; not something children are encouraged to do now in their spare time! I have to wonder, were we country children really so hard done by, stone picking, ragwort pulling, thistle cutting and potato picking?

This hedgerow takes you to the rear of Batchelors Farm with its windbreak of conifers. Another classic Isle of Wight farmhouse sympathetically modernised. Note the two little windows in the gable end either side of the chimney. The path sweeps around to the left of the farm to join its entrance drive at the front. The main barn has been converted; in place of the great wagon doors there are now long picture windows.

Continue down the farm entrance drive, this deposits you on Canteen Road, the name said to have been given during the First World War when, apparently, there were messing facilities here for German POW's. This road has fairly recently been widened and realigned. Across the road to the right is the start of footpath NC32. Take this path across the fields, surprisingly flat, for the Island. Skylarks seem more plentiful here. The establishment to the right front is Rill Farm, now a nursery and very much spruced up. The name is said to combine the growing of rye with a hill. The old farmhouse is just visible behind the new chalet bungalow. Across the field you come to an apparent fork; take the left fork, keeping the laurel hedge & fence to your right, until you meet the bridleway NC31 crossing ahead. Turn right onto the bridleway.

A dozen yards takes you to the original pathway junction and it will be noticed that the NC32 continues through the gap in the hedge to the left; it goes to Apse Manor. Although not the path to take on this occasion, a look through the gap reveals a large shed to the right front; most of the shed is now fairly new but this was an aircraft hanger when the field between here and Apse was an early civilian airfield. Known, from 1929, as Apse Airfield it has since been 'moved' twice and is now at Black Pan common, inland from Lake and Sandown and known as Lea.

To continue the walk, go straight ahead with the nursery and buildings on the right. The ruined lower courses of an old barn line part of the path by the fence. Past the farm the old farmhouse is seen to the right and at the same time you enter an ancient sunken pathway which in a rainy season can resemble the bed of a stream! As you climb, the reasonably flat ground of the old airfield to the left can be more fully appreciated. Badgers have been busy here.

When you emerge from the sunken section of path there are open fields to the right and a ditch and hedgerow to the left which could be the original Apse manorial boundary. Rill Cottage used to be here on the right, but it's long since been demolished. Continue straight ahead. To the right Stenbury Down comes into view with its Worsley monument and communication masts. Through the hedge to the left are glimpses of the St. Martin's and Shanklin Downs.

The next field boundary ahead is marked by a hedge coming in from the right, and the wreckage of various gateposts and past stiles in your path. **Immediately after this boundary PLUTO crossed the path**. A good marker remains here, one of the best preserved, if you can find it – a clue is the holly tree is growing through it! A short walk from here brings you to the main Godshill to Shanklin road. Go straight across when clear (Beware the fast traffic!).

The bridleway NC31, now marked 'To St. Martin's Down', continues on the other side. The house to the right is known, with good reason, as 'Island View'. At the end of that property a path is signed to the right that would take you down onto the old Shanklin to Ventnor railway track but your route goes straight ahead to the bridge over the old railway. This bridge, having had no real maintenance for some forty years, is a great tribute to its Victorian bricklayers; it spans one of the deepest cuttings on the Island.

Due to the trees the views from the bridge are limited. To the left Bembridge Down and Culver Cliff can be seen and to the right the ever present Stenbury Down. If you haven't a head for heights don't look over the parapet in the centre. The railway was in use for just five months short of one hundred years from September 1866.

From the bridge a gate takes you onto what is often called a rew, a path or track following a row of something, in this case a hedgerow. About seventy-five yards past the railway bridge there used to be a junction of bridle ways;

the one you are on being crossed by one connecting Shanklin with Whitely Bank. Before the road you have just crossed was made in 1885 this was the only direct route between the two. Continue with the hedge on the left and just soak up the scenery. Soon after the gate glimpses of Godshill Church and the Freemantle Gate can be seen away to the right; and a huge expanse of the Island is laid out behind you. The hedge on the left is almost certainly a continuation of the old manorial boundary.

As you rise, the tree lined cliff top of St. Martin's down becomes apparent. The geometrical shapes in the centre below the top are reservoirs. Why St. Martin's? It is alleged to be a corruption of 'Smerdone', but could it have had something to do with the very ancient family of St. Martin who were Lords of the Manor of Alvington? Or would that be too far from home?

Another bridle gate takes you into a fenced off section of the path. About halfway up this part of the path is a shallow pit in the field bordering the path; it looks as though it could have been a dewpond, or was it caused by a stray bomb from the attacks on Ventnor?

At the end of this path, near the trees, no fewer than seven footpaths meet. There used to be a cast iron footpath signpost here cast by Wheeler and Hursts in Newport, but this has now been discarded, (It's still in the hedge) and three separate posts, which will corrode, put in its place. At the corner of the field turn immediately right onto the footpath V50 orV34 to Winstone, as shown by the first of the signposts, the other two are through the gap ahead.

Having turned to the right the official route of V50 goes immediately back into the field via a stile obvious on the right. In the field the path is constrained by a fence; at the time of writing, there is major redesign of footpaths going on at the end, and V50 rejoins V34. There used to be a difficult bog here; this has been drained, and a nice pond at the bottom sits between you and the next gate – the path goes to the right of this, the drains are to the left.

When you reach the stile, you will find it superfluous that it has a gate on the far side which is the actual boundary with the next field. At this point, through the gate, NC43 goes to the right but your path, still V50, continues across the field; slightly to the left. In good visibility, head for Godshill church tower or Chillerton TV mast. After a few yards there is a good view, through its trees, of Appuldurcombe House to the left.

Go through the field gate in the dip, and then head slightly right across the shoulder of the rise ahead to the stile which soon comes into view. Various local farms are in view from the stile. Yard Farm is tucked down on the left. It was given to Quarr Abbey in 1188. Across the valley is Appuldurcombe Farm, the home farm of that mansion. Redhill and Park Hill farms follow this to the right and closer on the right, is the Donkey Sanctuary.

Winstone Farm your next destination is too close under the hill, to the right front, to be seen. There were brick-works in the fields away to the right; substantial enough to have their own spur from the railway.

From the stile, aim for the electricity transmission pole diagonally across the field to the right, after a short distance you will see a gate in the corner of the field. Through the gate on the left an old, twisted and sunken path leads down towards Winstone Farm. The path ends at the disused Shanklin to Ventnor railway track; now used by walkers and cyclists. There was always a crossing here. Straight across the track a stile takes you across a small paddock, the other side of which is another stile onto Winstone Farm's access lane.

Winstone, at the time of writing, is, I am pleased to say, still farmed in the traditional manner as a dairy farm; there are so few left. Historically, as a holding, it is of similar age to Yard. In the farm drive continue down to the road. This is the main road from Whiteley Bank to Wroxall and Ventnor. The path continues across the road twenty yards to the right. This is now Bridleway NC43 and it runs straight, between tall hedges, into the main yard of Lower Winstone Farm, now a Donkey Sanctuary. (What is now the bridleway used to be the main entrance.)

Perhaps a word here on donkeys. Often a much misused animal; but as G.K.Chesterton wrote; for the donkey :-

> 'Fools! For I also had my hour;
> One far fierce hour and sweet:
> There was a shout about my ears,
> And palms before my feet'.

(I always look to see how well defined the cross is on their backs)

When General Allenby was about to make his triumphal entry into Jerusalem after capturing it from the Turks in December 1917; he apparently asked his staff to plan the event. When told of the arrangements, he queried why, as a cavalryman, he wasn't to ride through the gate on his charger. He was reminded that perhaps a greater man had ridden through that gate some 1900 years previously, and it might be inappropriate to appear to re-enact that event. The General walked through.

The path now goes straight ahead between the farmhouse and the barn. A standard WW11 air raid shelter survived in this gap, until 2005. Both buildings are old. It used to be known as Lower Winstone Farm, but on a map of 1764 it is referred to as Wiston Mill and Wynsone Myll in 1586.

After having passed between the buildings a look at the land contours to the left suggests that part of the existing house could have been the mill. This assumes that the water came from a dam on the main stream further up the valley There is definitely a man-made terrace running directly to the buildings from the south which would appear to have contained a leat to bring water to the wheel. It is likely that one of Wroxall's Domesday Book mills was sited here although that would probably have been down on the actual stream. Old maps also show a sheep wash between the buildings and the main stream, presumably utilising the millrace. (A path GL39 goes south along that terrace to Redhill Farm.)

Cross the stream at the bridge; you are now at a footpath junction, GL36 goes straight up ahead and your path GL34 turns immediately to the right. Note the paths are now GL's for Godshill instead of NC for Newchurch as the stream marks the boundary of the parishes. The stream, which originates from springs in the valley just west of Wroxall, does not appear to have been given a generally used name. It has been known as French Myll Brooke, Wroxall Stream and Whetley Stream at various times.

The path, starting with a stile, now borders the stream. Initially it is a veritable marsh and quite difficult going, in April there is a fine crop of kingcups. A double stile with a footbridge will bring you to somewhat firmer ground. The way is now through a streamside meadow but the path is not well defined so, where possible, just keep to the stream on the right and follow its twists and turns. The meadow ends in a narrow neck of land with a sewage-settling pond in the left-hand corner, you may get a whiff. About twenty yards from the end of this narrow neck, on the right hand side the path turns right over a stile and footbridge crossing the stream. This point marks the upper end of the millpond for French Mill. The other end of the footbridge has another stile after which the path turns left. The sunken area to the left, containing the stream, which was the millpond, has been turned into an attractive wild garden, often a blaze of colour. The buildings of French Mill, now a private house, soon come into view. It would appear that the owner kindly keeps the path tidy on this side as well and is probably responsible for the plantation of saplings.

Why French mill? Obviously French owned or tenanted. Appulducombe Priory was founded in about 1102 under the jurisdiction of the Abbey of Monteburg in Normandy; it was known to own or hold property in the Wroxall and Sandford areas. Domesday records two mills in both these areas. It would seem likely that sometime between 1102 and 1414 the priory was owned and/or ran at least one of these mills; hence possibly French Mill! The name appears to go back to at least 1316. The stile at the end of this section by the bridge, (now much widened on the northern side), deposits you back at your starting point on the Godshill to Shanklin road.

**Walk 13**

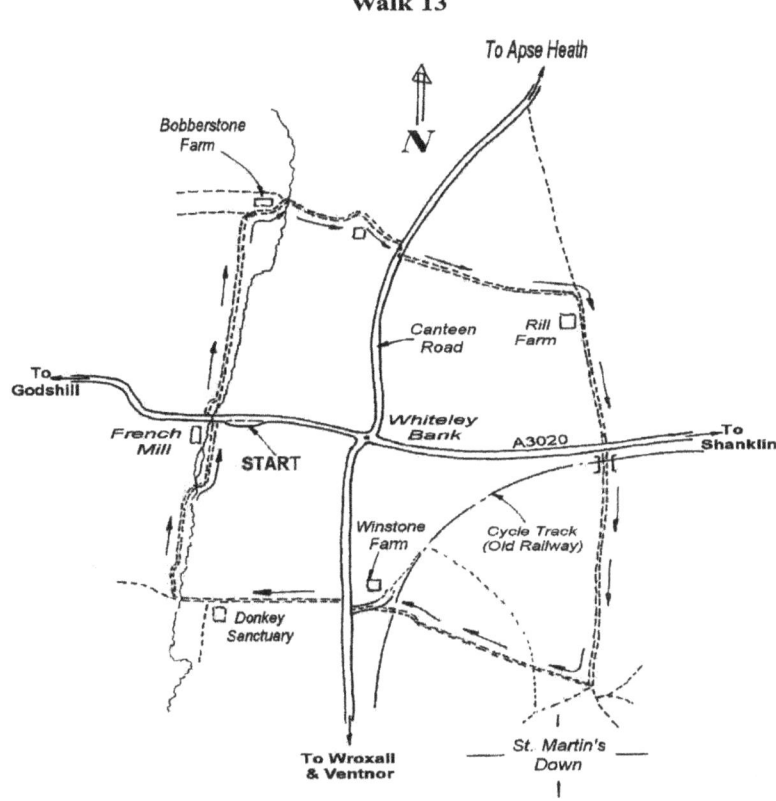

*Not to scale*

# 14. THE START OF THE OLD TRAIL

*(The 'Tennyson' that is)*      *(About 2½ miles.)*

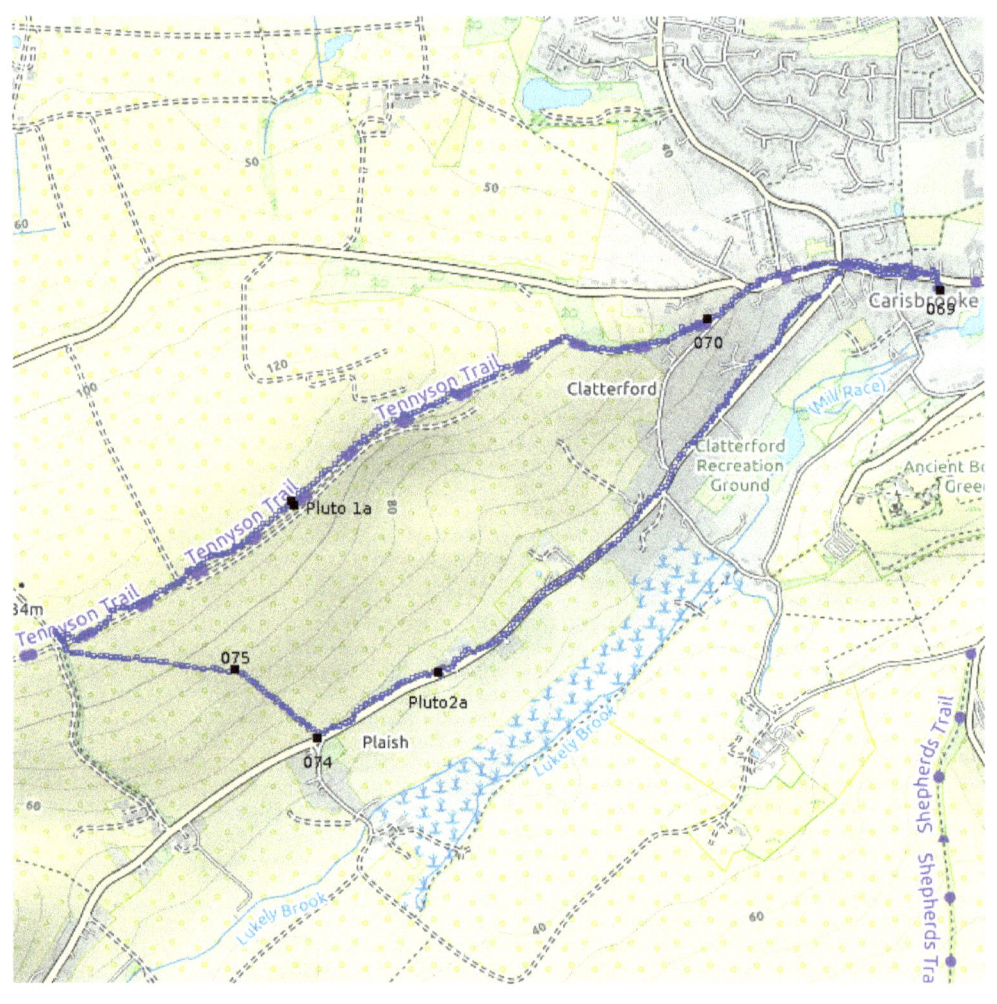

The Tennyson trail, running from Carisbrooke to Freshwater, and on to the Needles, was one of the firsts of the Islands 'long' distance trails to be formalised and given a name, in 1968. The first leaflet detailing the walk was issued in May 1970. It had of course been an established way ever since our forebears walked on the high ground instead of in the valleys. Sections of it remain classified as Ancient Roads. The trail is one of the finest walks on the island for vistas and its lack of modern interference; only three roads need to be crossed in its fourteen miles, quite remarkable for the Island. It is well known and described elsewhere. This chapter will only cover the first mile or so.

Make a start, uphill, from the car park in the centre of Carisbrooke. This is charged for parking 7 days per week, and has toilets. As you pass the first building on the left look for the new milestone on the inside edge of the pavement at the end of that property; this is the 'one mile from Newport' milestone. This High Street is full of old and interesting buildings; can you spot the 'mathematical' bricks from the real ones? They are used on some of the houses on the left.

Obvious on the right is the island's largest mediaeval church, St. Mary's; once part of a long demolished priory. A little further, on the same side, just outside the lynch gate is a pair of cottages made out of what was once the Cutter's Arms pub. One wonders what they were Cutters of, but Kevin Mitchell in his book on Newport Pubs tells us that it had previously had its proper title 'The Gelder's Arms' so now we know what they were cutters of.

Go straight over the crossroads, continuing up the hill. The Waverly Hotel on the left was, originally, nothing to do with the paddle steamer of the same name but was given its name by the Wavell family who built it. They were wine merchants and publicans of Newport. Go past School Lane on the right and Pit Terrace, built down in the old marl pit, on the left. Opposite the converted chapel on the right, take the left fork into Nodgham Lane. (The main road continues up over the shoulder of Alvington Down to Calbourne. A wonderful description of a wagon taking this road, up out of Carisbrooke, is to be found in 'The Silence of Dean Maitland' written by Maxwell Gray, a Newport doctor's daughter.)

Only continue up Nodgham Lane for about one hundred yards before taking a right hand fork onto an unmade, chalk based, track which is, now, the start of the ancient downland way. (The N123 Byway). It was, of course, once used far more than today, even allowing for our walkers and off road cyclists. This initial part of the trail was particularly well used in the late eighteenth and early nineteenth centuries to take the local populace up to the races.

The first quarter of a mile is fairly hard going for casual walkers but the rewards make it worthwhile. In the first wooded section you unknowingly pass Alvington reservoirs numbers one and two, high up to the right. More obvious, also on the right at the end of the wooded area, are some of the, ever growing plague of phone masts. This first part of the down is known as Alvington from the ancient manor of that name; its central farmstead and manor house is situated just north of the Calbourne road.

When you break out from the high hedges and banks lining this first section you will find yourself in an elongated, so far unspoilt, piece of meadowland. Apart from an occasional clean up this strip of meadow is usually left untouched and is a haven for wild flowers and butterflies. Between the wars, when people had more time and fewer distractions, it was a favourite spot for Carisbrooke families to picnic. Given the right weather, the views from the middle of this section are extensive.

For those who do not know the topography well, it is worth 'pointing' out the features. Listing them from just north of east, on the horizon, to the left, there is Staplers Heath and Lynn Common with the Ashey sea-mark peeping up between that and Arreton Down with its scar of chalk. Then follows St. George's Down, and the high ground behind Sandown and Shanklin. Shanklin, St. Martin's, St. Boniface, and Wroxall Downs come next, all rolled into one large lump. After this is Stenbury, slightly nearer, with Rew Down to its rear followed, on the far horizon, by the tops of the inland cliffs above the Undercliff, terminating in St. Catherine's of which only the top of its mast can be seen.

In from the horizon, in front of St. Boniface from here, is the initially tree lined, high ground above Whitcombe and Gatcombe, and a corner of Chillerton Down. This is followed by another great mass of Garston's, Dukeum, Westridge and Northcourt Downs. At this point the road to Shorwell can be seen sneaking over the divide. This is followed by the Shorwell, Rowborough, and Idlecombe Downs. Ahead, having now taken the view right around to the south-west, are the trees capping Brighstone Down, which lies on the route of the Tennyson Trail.

Spread out below is the whole of the Bowcombe Valley through which the Newport to Shorwell road runs. A look back a little further on gives a wonderful view of Carisbrooke Castle.

The beauty of the Island was appreciated by William Lisle Bowles; when viewing it from Southampton Water he wrote:-

Vectis There

*That slopes its greensward to the lambent wave,*

*And shows through softest haze its woods and domes,*

*With grey St. Catherines creeping to the sky,*

*Seems like a modest maid, who charms the more*

*Concealing half her beauties.*

From 'Cadland, Southampton River' by W.L. Bowles

After a couple of hundred yards the path closes in again, proceeding gently uphill. A glance to the higher ground on the right will reveal the rather geometric shape of the relatively new Alvington Number Three reservoir.

Very occasionally the hedges of saplings of this narrower section are cut back (or, in these days, massacred by flailing machinery) but there are many years when this section becomes a virtual tunnel. As befits an old highway there is a great variety of saplings and other plants in the hedgerows, although hazel, oak and field maple seem dominant along with the ubiquitous, chalk loving, clematis or 'old man's beard'.

**About one fifth of a mile (or 350 yards.) from the end of the meadow section, and as the path opens out a little, near some taller, mature trees PLUTO crossed the path**. It is difficult to give before and after pointers here; but there are remains of markers on both sides of the track, one side considerably better than the other.

The track now flattens out and although it runs nearer the top of the down it keeps just far enough below the crest to maintain, very sensibly, protection from the north. People of old must have been very tough but were obviously not averse to avoiding unnecessary hardship. At this higher harder level, it is noticeable that the hedges have now become more prominently blackthorn.

After a fair distance you come to a crossing of the ways. The N126 is the path to take. A fingerpost shows all of the other options.

The N125 goes over the top of the down, to the Calbourne road and possibly a pint at the Blacksmiths Arms opposite. (However walking the main road either side of the Blacksmiths Arms is not a good idea). The N127 goes down to the Shorwell road and the N128 continues as the Tennyson trail across what is now Monkham Down along what is alleged to be a section of Roman road. It is known as Monkham simply because monks owned it as a detached part of Rowborough Farm, when that was an abbey grange.

It is on this hilltop that the old Monkham Racecourse was sited. The depression, in the field, to the right front adjacent to the junction of the paths used to be known as Gypsy's Hollow; probably stemming from horse racing days. Old maps have it as a quarry. My suspicion is that it was originally a dew pond; although not in living memory.

If a longer walk is wanted there are obviously many options here to make your own itinerary. However, as this is basically a PLUTO walk I suggest that you take that path, the N126, going diagonally back across the upper part of the field, almost a hairpin. It is not very visible.

This path across the fields gives great views of the Bowcombe valley and its variety of fields both arable and pasture. There are hares in this field. Back to the right is the hamlet of Bowcombe itself and more immediately to the right is Plaish. It is this valley that provided Newport's first, organised water supply by the tapping of the Lukley stream's headwaters and tunnelling it into Newport. (There is a marvellous description of this in Bill Shepard's book 'Newport Remembered') The group of tall poplars between Bowcombe and Plaish hide the pumping station which, later, pumped water up to the Alvington reservoirs.

A double stile at the end of this field is a convenient place to rest and soak up the surroundings. Straight across the valley is Froglands Farm and, obvious, a little to the left, is the castle, with Mountjoy hill behind it. Not 'Mount Joy' as on the modern maps; it was named after Mountjoy Blunt, who was made Earl of Newport in 1628. After 1858 much of it became a cemetery!

In the next field the path goes more steeply down to the Shorwell road, where it emerges opposite the old smithy at Plaish. Turn left back toward Carisbrooke. Not an inspiring piece of road but having been widened it does, at this point, have a pavement on the left. **As you approach Goldings Farm Cottages on the left, PLUTO crossed your path.**

Fortunately, the road was widened before WW II so, at the time of writing, there are some remains of markers on both sides of the road. **PLUTO actually went under the end of the farmhouse garden.**

Soon after this the pavement runs out but at least there is a verge to walk on, or at least to take refuge on. It is not an ideal walk back into Carisbrooke; mainly due to traffic, and apart from the castle not a lot of interest. All three of the farms on route have been 'residentialised' One often wonders who owns and farms the land!

Bowcombe Chapel is soon passed on the right. This chapel may indeed be nearer the original site of Christian worship locally than Carisbrooke church itself. The church in Saxon times is referred to as being at Bowcombe. Nothing definite is known; it may be that a church was where it now is and have been called Bowcombe merely because it was in the 'Bowcombe Hundred. I believe this theory finds more favour.

In the fields opposite Bowcombe Barn farm, the next one on the left, there are the remains of a Romano-British villa; between the road and the Lukley stream, on the further side of the hedge which parallels the pipeline. This could support the theory that the area was more of a centre in the past but this is rather offset by the fact that there was another villa in what is now the garden of Carisbrooke vicarage!

Where the first houses appear, later, also on the right, it will be noticed that the 'Rat Trap' brick wall fronting the road is much older than the buildings behind it. This is because it was previously the site of an older, much more substantial, dwelling known as 'Castlehurst' which was demolished after a fire. (I think, in the 1970s). NOTE The folk now living there say they moved in, in 1958 – also they remember the pipeline scar showing in chalk when the field was ploughed.

The next turning on the right is Clatterford Shute, which may be a preferable, quieter, way to return. A left turn after crossing the ford at the bottom would take you into Millers Lane which would take you back to the centre of Carisbrooke. Almost opposite Clatterford Shute, on the left, is Nodgham Lane. It is from the other end of this lane that you started on the Tennyson Trail earlier.

Ahead is Clatterford Road which leads directly back to Carisbrooke. About halfway down this road on the right sat the old Infants School of 1889, long since unused as such, it has also been the County Archaeological Centre.

Sadly, at the time of this edition, it is slated for demolition with permission for 2 houses. The housing on either side of the road is typical of the ribbon development between the wars, with some more recent infilling. Much of the land to the left between Nodgham Lane and Clatterford Road belonged to the large Italianate style house which can be seen in the centre of that area. (Known in the 1880s, not surprisingly, as 'Castle View') One of its original boundary stones, looking remarkably like a milestone, can be seen opposite Pallissy House, the large building about three quarters of the way down the road on the right.

Also on the right near the end of the road is the tall Victorian vicarage which has the remains of the aforementioned Roman villa in its garden. There has been talk in recent years of conducting a 'modern' excavation and putting it on show. At the time of writing nothing has come of this and it is known to have deteriorated rather badly since its discovery in 1859. At the crossroads turn right to return to your starting place, unless you require refreshment in the Waverly Public House, to your left!

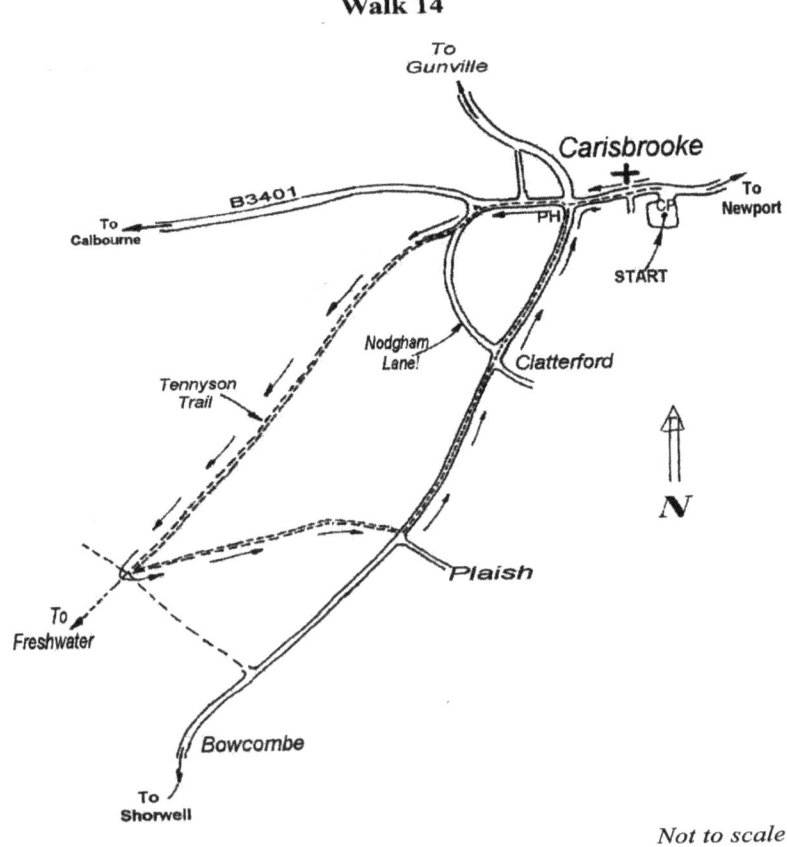

**Walk 14**

*Not to scale*

# 15. A WALK IN CENTRAL FARMLAND

*(Starting from 'Merstone Junction')*   *(About 3¾ miles.)*

This is a walk in the farming countryside which many will enjoy even though it has nothing particularly dramatic or historically fascinating about it. Perhaps it's the sort of walk when a partner would be more than usually welcome to talk as well as walk!

However, to those interested in topography it is a unique part of the Island. Here there has been, and perhaps still is, a battle between the Island's two major rivers with regard to their watersheds. Both the River Medina and the Eastern Yar have tried to claim the area for their own. Initially the walk is in very flat country and the dilemma of the drainage routes are obvious although here the Medina generally wins. Later it becomes more complex.

If the watersheds of rivers are not exactly one of your great interests then please skip the bits where I drip (!) on about it and just use the directions for a good walk.

Starting from Merstone Junction, or should I say, where the Newport to Sandown cycle track crosses Merstone Lane, (it's easy to park here). We will start off north westerly toward Newport. (Merestone in the Domesday Book i.e. marsh with a farm).

A few words here about Merstone Junction. This was where the Ventnor West line parted company with that from Newport to Sandown, and was one of only two Island stations with a subway, initially thought necessary to access the central platform from the road. It was parallel with the road about where the entrance is now; unfortunately, it filled with water at the first sign of rain, (that drainage problem again), and eventually was only used as a water tank to replenish the engines.

So, after taking in the old platform and the Scots firs to the left, you head off, the gravel workings on St Georges Down are prominent to the right front. After some 340 yards a footpath crosses the old railway, go left here, on a plank or two over the ditch, which flows NW, (to Blackwater and the Medina).

You are now on FP SA33, the path goes slightly to the left across the field to, about, the centre of the tree-lined boundary opposite; when you get there you find that it is actually a corner and the path now follows the boundary, bearing slightly to the right. The view from here is mainly of arable land although to the right one can see Alvington Down, the high ground above Whitcombe, Garstons Down and New Barn Down with Tolt Copse clinging to its side. Locally, just ahead, part of Pagham Farm can be seen.

The field boundary now on your left is obviously very old with ash, oak, hawthorn, elm (some quite healthy), blackthorn and many other species; even some very mature black poplars.

> *'Give me a land of boughs in leaf,*
>
> *A land of trees that stand,*
>
> *Where trees are fallen, there is grief,*
>
> *I love no leafless land'*
>
> A. E. Housman

At the end of the field the view is reduced to crops but looking back you have in view the whole of the east end of St. Georges Down and then Arreton Down with, initially, its line of chalk workings.

The path now crosses a deep field drain, part of the watershed battle. This one again flows NW to the Blackwater (some say Birchmore) Stream which feeds the Medina. You bear slightly left over the bridge, now joining the SA 36 bridleway. After about 80 yards the track turns left and the FP SA33 goes straight ahead across a field aiming for the right hand end of the trees to the right of the farm. However, the field usually has crops and it is more interesting to continue along the track between the trees to the left, heading SSE.

This can be a quagmire in a rainy season and well illustrates the fact that the water could go either way. However, if it makes it to the very deep ditch on the right it goes Medina's way. Away to the left, almost parallel, is Merstone Lane which, as it has been hardened up over the centuries, has become something of a causeway, preventing drainage here reaching the East Yar. This short stretch of track also has some fine trees, note especially the unusually tall silver birches.

At the end of the track turn right, the ditch is bridged here, and take the track up towards Pagham Farm, the SA35, this is an absolutely classic country track and a bit of nature at its best, despite the intensive agriculture either side.

When you reach some farm buildings on the right, pause to take in the view to the left, from the two houses on Merstone Lane clockwise to Stenbury Down with its Worsley Monument and communications masts. The land here is now sloping away to the south and east but due to Merstone Lane any water still ends up in the Medina! Carry on past further buildings on the right, loose boxes and pens sadly only used now as storage due to the change here to arable only. At the end of the track turn right into the farm, noting the changes over the years to the barn on the left. (When you turn right here you leave SA35 but in a few yards rejoin SA33 when turning left).

Pagham is obviously a very old settlement and in times past was an outlier of Merstone Manor known as South Merstone. However somewhere along the line it is alleged that a family called De Pagham, from Sussex, became involved with the ownership, hence the name.

The farmhouse has, no doubt, been rebuilt a few times through perhaps 1000 years, but this one looks quite old enough. (17c). As you turn up past the left hand end of the house please note the lower courses in stone followed by old brickwork, and the substantial buttresses in addition to wall retaining irons at the end of the tie bars. The tie bars will be running under the bedroom floors holding the end walls together. Also, as you turn, note the magnificent example of the multipurpose barn on the left. It has seen better days but what character it has! I hope it's still there when you pass by.

Continue up the now gravelled, lane which is the entrance drive to the farm, past another building on the left which is newer but not that new as will be seen from the bond used in the brickwork. This is none of the more recognisable bonds; my guess would be Victorian, based on its condition, but it would be interesting to get an expert opinion. (Strangely it appears to be Sussex Bond but it can't date back to the De Paghams!).

The lane now does an S bend, ignore the path back to the left, you will come back to that later (That path is part of SA35 which runs with SA33 here). Carry on through the trees for about 100 yards; this section of the lane forms a causeway above the low marshy ground in the wood on the left.

At the end of the trees bear right through a field entrance which is the start of a field track. After about 15 yards bear slightly left; do not follow the track around to the right, the footpath goes across the field toward the line of trees ahead, aim about 30 yards to the left of the right- hand end of the trees.

**About one third of the way across this field PLUTO crossed the path;** there would have been no markers here as there was no hedgerow.

When you reach the trees and the field boundary, turn left. The FP used to go straight on here and was the path from Rookley village to Pagham but due to a modern diversion it now goes down the hill to the Rookley to Godshill road where the SA33 ends. At the road turn left, preferably walking on the verge (sometimes, depending on crops, it is possible to walk inside the boundary of the field opposite; but that would entail crossing the road later where it's difficult to be seen). Continue up the road in the Godshill direction but only for the length of the field on the left. At this point you arrive at the entrance to Pagham Lane (Bridleway SA35).

Turn left down the lane, leaving the cottages on the right. The lane goes down to a noticeable dip where an obvious, but miniature, valley comes down from the left and drains into a wooded boggy area to the right; where this water goes is one of the puzzles. **Ignoring a few trees in the hedgerow to the right but at the start of the actual wood on the right PLUTO crossed the path.** There is a rather decrepit marker here; it's difficult to find.

A few yards further brings you to the point where you left this lane a little while ago, (where SA35 and SA33 run together), retrace your previous steps through the trees to the right but in the S bend take that FP, now on the right, which remains SA35.

The wood to the right is now lined with ancient oaks and the roofs of Pagham Farm can be seen again away to your left front. Toward the end of this short stretch of path stands a wonderful old ash tree, with all its battle scars; the dead bits providing homes to woodpeckers. Then more massive oaks and when you clear these there is, straight ahead, a view of the knoll above Redway and behind it, Brading Down.

At the end of this path turn right, away from the farm, you will now be on the SA32 (do not continue on SA35 which goes left then right). Continue down a farm track for about 150 yards where you have to turn right again. (The track used to go straight on and join up with Merstone Lane!) As you turn note again the deep ditch to the left still linking up with the Medina watershed.

The path now takes you back through the trees and again it is a causeway build up through hundreds of years over the boggy ground. The almost insignificant hump in this short stretch is, I believe, actually a divide which separates the water drainage here between the Medina and the Eastern Yar.

On emerging from this little bit of woodland you need to bear left along the edge of the remaining trees; at the end of the trees the countryside opens up and there are good views of Stenbury, St. Boniface, Wroxall, and Shanklin Downs ahead. The path now bears right across a boundary between two fields. **Shortly after leaving the trees PLUTO crossed the path;** there is no trace of a marker; perhaps there was never a hedge. The next little 'wooded' area you come to on the right was the site of a wooden bungalow, and its garden, which until some forty-five years ago was a family home. Sadly, all that is now left are the remains of the brick chimney-breast. After this the track takes you down to the Rookley to Godshill road.

At the road turn left, onto the verge, and proceed to the next turning on the left; this is Bohemia Corner, named after the farm on the right. Why Bohemia? No one knows, there is a suggestion that it has some connection with a tenement at Roud, but the deeds quoted are dated 1730 whereas the deeds of Bohemia farm, quoting the name, go back to 'the 1600s'. It may be a fanciful notion but perhaps some crusader named it after returning via Germany!

Turn left along Merstone Lane with the RSPCA dogs home on the right, carry on along the lane, being very careful of the two-way traffic in this narrow lane. (Drivers always seem startled; someone actually walking! They just don't consider or allow for it!)

**PLUTO crossed the road at about the end of the dog's home grounds**; some 50 yards before the lane bends to the left. There is one of the best-preserved markers here; but only on the south side of the road. I am sure the lane has been widened since 1944. Cross to the right hand side of the road before the bend as soon as it is practical (ie. as soon as there is just enough room on that side to evade the traffic) At the apex of the bend turn right into the bridleway SA22 to Budbridge Manor. This bend bridges a considerable stream running from the area you have been walking in but now definitely flowing SE into the Eastern Yar.

As you start on the bridleway note the rather idyllic setting of Bohemia cottage to the right and ignore the FP SA46 to Little Kennerly also on the right. Continue straight ahead; there is a wonderful stretch of marshland down on the right and across it, in the trees, is Little Kennerly (Kenewardle, apparently in 1202). Two footpaths, the SA46 you have just passed and the SA47, marked by a stile just ahead, cross this marsh but wellies are advisable in dry seasons and swimwear more suitable in the wet! The SA47 was the way in which people got from Little to Great Kennerly; the access track to which was track you are now on.

Nothing remains now to show the exact site of Great Kennerly, but it was somewhere quite close to the 'dog leg' in the track you now come to. A little further on, the track crosses the old route of the Ventnor West railway, clearly shown by the hump in the track. It is quite likely that the building of the railway removed the last traces of Great Kennerly. (Or is there more of a story to be uncovered here?)

Continue past the old railway, the route of which is completely overgrown here. Contrary to some figures that have been given, only around 38% of the Island's unused rail system has been secured for use as cycle ways and public paths. Undoubtedly a good effort by recent councils, but what a dreadful mistake to have allowed any of the old permanent ways back into private hands.

Aim to the left of the trees ahead (the path is not always obvious after ploughing), You are now in the 'nursery land' surrounding Budbridge Manor. The trees act as windbreaks for the glasshouses and, I think, help preserve some semblance of the countryside. At the end of the trees turn left, leaving the manor away to the right, you depart here from SA22 onto Budbridge Lane.

This lane up from Budbridge is also on the route of another PLUTO walk (see Walk No 8) but it is preferable to use it rather than to have tried to walk further on Merstone Lane, which is dangerous in places. I will not, therefore, dwell unduly here suffice to say, perhaps again, I just love Little Budbridge, which you come to a little way up the lane; the farm is partly in ruins which adds to its romance and the farmhouse is a real classic.

Although you turn left just before the top of the next rise it is worth going a little further, to the top which is in fact an old rail bridge presently taking the lane over the Newport to Sandown cycle track. It had been filled in underneath but they remade the cutting when the cycle track was opened in 2003. The view from here is surprisingly extensive; from the knoll above Redway to the right front all the way around anticlockwise to the west, to the Hoy Monument behind St. Catherines and, without the foliage in the winter, even to St. Boniface in the SE.

However, it is necessary to go back to that left turn (or is it right now!) and take the path north-westward alongside the cycle track. This is Bridleway SA54 (The cycle track is SA4) which takes you to Merstone Lane. At the end of the field you will see a double hedge going away to the left (SE) which is again the line of the old Ventnor West railway, as it left Merstone Junction. The bungalows on the right have been built since the closure of the railway; the FP SA4 goes behind them to the road, but continuing in front of them takes you to the road, across which is your starting point.

Rather a contorted route, **having had your path crossed by PLUTO four times!**

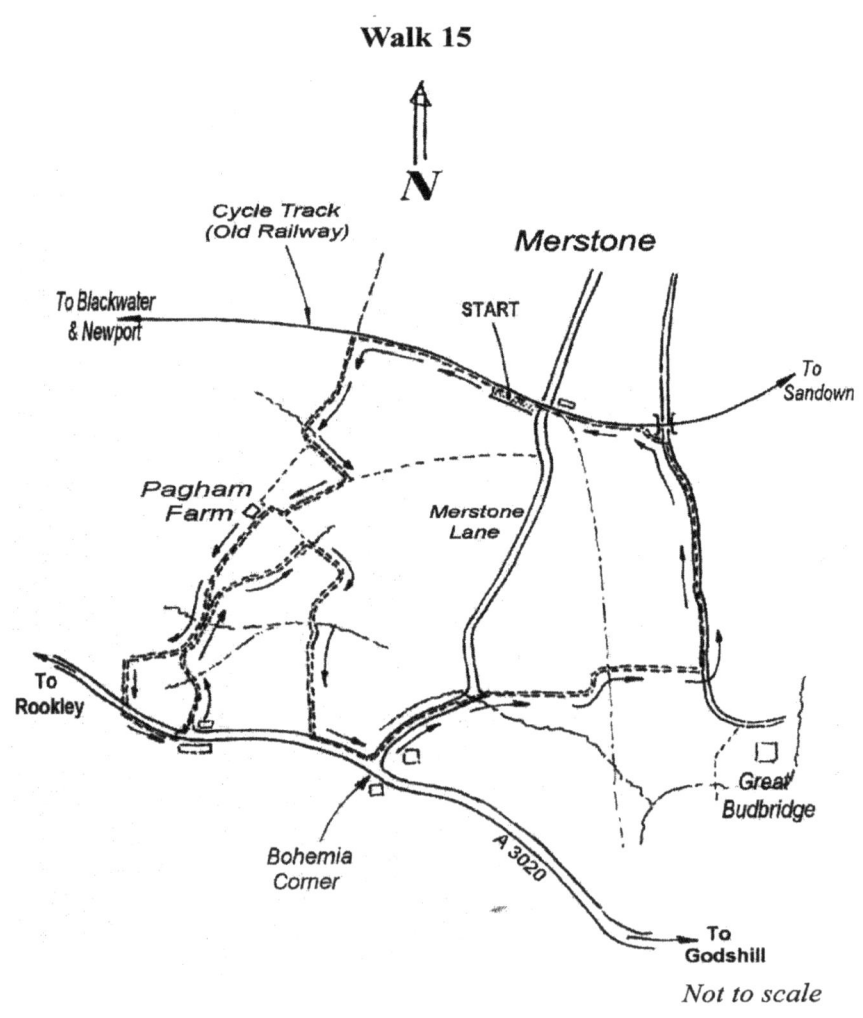

*Not to scale*

## 16. BACK 'O' WHITECROFT BOOTS & HILLS

*(Another extended ramble south of Newport)*   *(About 5½ mls.)*

Well that's what we have always called this walk, which takes in a very central part of the Island south of Newport; it can be varied or extended to suit the weather or the mood.

A convenient starting point, in order to 'take in' part of older Newport is the crossroads adjacent to St. John's Church. Diagonally opposite the Church is the old Mechanics Institute building, sponsored by the Seely family, which later became the County Seely Library and is now part of the school complex. On the south east corner stands a cottage, the older part of which served a toll gate although now called St. John's Cottage. The church itself, originally a 'chapel of ease' for Carisbrooke, predates the rebuilding of the Newport parish church, having been erected in 1837.

Your route, St John's Road, goes south with what is known as 'The South Mall' on the left hand side. Over the years there has been much 'infilling' of housing of various dates and styles along this road but some of the older ones are worth pausing for. A short distance up the hill on the left is the cream coloured 'Forest Villa', the home, up to his death in 1909, of one of Newport's famous hymn writers, Mr. Albert Midlane. A little further, on the same side, though hidden by high walls, is 'Pit Villa' one of the first houses build on this land which used to be known as 'Trattles Butt'; a place, apparently, where the youth of the town used to 'resolve their differences'!

Toward the top of this 'Mall' there is a wonderful old terrace which at first glance appears uniform, but a closer look at the houses shows surprising differences. Some, unusually in Newport, have basements. A little further, on the other side of the road, there are very stately Victorian 'semis' of three and four stories and a very lofty Chilean pine. The houses next on the right have been developed on the site of Shide Cross Farm.

The road then drops slightly to a crossroads known as Shide Cross which is now a mini roundabout! St. Nicholas Place, to the right immediately before the crossroads, used to be a villa with fine brickwork.

Your route continues straight across into Watergate Road. On the right hand corner is 'Elmslee' which has recently had a vertical extension worthy of an award for the way in which it has been integrated with the original. A little way further up, on the same side, we see Watergate Cottages, which used to be known as Watergate Farm (there's another later!) with its 'interesting' sideways extensions. Continuing up the hill there are neat little Victorian villas to the left and more modern housing to the right.

After a quarter of a mile or so the housing comes to an end and you can start to see the countryside emerging. To the left, at the end of the houses, there used to be an old five-barred gate on which one could lean and admire the view to the south-east. Unfortunately, a few years ago, this was replaced with a high, galvanised steel monstrosity, which although possible to see through, is quite out of keeping with its surroundings. In any case the view is now obscured by bushes and the back of the café.

A short distance on, however, the entrance to the pet centre, on the left, gives a clearer view and the café offers refreshments combined with a panorama; it is perhaps slightly spoilt by 'activity' in the foreground but nevertheless worthy of contemplation.

The view is generally south over the Medina river valley. Progressing round from the East, your left, you have St. George's Down, in its best light from here as the extensive gravel workings are not really noticeable on the slope of the down. Details of these later. Coming round you have the St. Boniface and Wroxall downs in the south-east then Stenbury Down. The lakes in the valley used to be quite picturesque, they are entirely man made in recent years, however not now visible. Further round still, you can glimpse some of the roofs of Rookley, in the saddle between the hills. Due south, somewhat nearer, is Bunkers Hill, just west of Rookley; after that Great Down, with its topping of Scots firs obliterates everything until, in the winter, one can just see the woodland behind Great Whitcombe Manor to the south west.

Having absorbed the scenery, continue down the lane, passing the now (2003) rather derelict nursery on the right which will, hopefully, have been tidied up by the time this is read. Next on the right is a rather nice Victorian vicarage which was actually the second one built for St. John's. Opposite are the modern kennels with an accompanying house which is nice enough in itself and now blends reasonably into the countryside but why permission was given to allow it to be built in such an obviously 'green' spot puzzles me.

A few yards further, on the right, is the drive into Newclose House; a grand house in a very attractive south facing setting, albeit now rather 'submerged' under its own trees. It is now partitioned.

Continue down the often wet lane, (there are springs easing from the bank here) you soon reach the 'Watergate' itself which is, now, in reality just the bridge over the small stream. There was once a tollgate here and the tolls were collected at a booth; the wall of which is said to be incorporated into the boundary wall of the property to your left, certainly there are bricked up 'windows' still visible but was this the toll booth?

'But never more in the dim moonbeam

Than a cloak and a plume and the silver gleam

Of passing spurs in the night can he see,

For the toll-gate's gone and the road is free'.

From 'The Toll-Gate House' by John Drinkwater

You appear to be at a junction of five ways here, but it's really only three as the track on our right, now serving two or three houses, was the secondary entrance to Newclose House; next right is Nunnery Lane, then the new entrance to Watergate Farm; your route, however, continues over the bridge on the metalled lane. The bridge has never looked the same since a local notary accidentally 'removed' the old brick parapet, however the classic brick arch beneath the road is probably the original; before that no doubt it was a ford.

The area just over the bridge, to the right, has been cleaned up in recent years in concert with the blocking off of the original entrance to the farm. In February especially it has a wonderful array of snowdrops amongst the old elm stumps. Watergate Farm is now immediately on your right.

Turn right into the bridleway (N110 to Whitecroft) on the right, immediately past the farm. The bridleway, although recently improved, can be very muddy, depending on the amount of recent horse traffic, and sometimes it is a veritable stream; but it is a path not to be missed.

The path rising past the farm becomes classically sunken and is quite narrow, never a cart track this one, with ancient hedgerows to both sides, including a surprising number of elms; a naturalist's delight. Near the top of this stretch the path turns sharply to the right and skirts around a site which may once have been a gamekeeper's cottage; there used to be traces of gun dog kennels in the trees to the right. This is being redeveloped at the time of writing (2019).

At the top end of the site, the path emerges onto a wide track, turn right on this and continue up through what the children have always called 'spooky woods'. There are some lovely old trees here despite there having been some casualties during the fierce gales of 1987 and 90; note the particularly magnificent beech, to the right. The track bears slowly south again as it rises through a short steep cutting in the actual bedrock.

At the top, the track breaks clear of the trees and you are, quite suddenly, on top of the highest point around; albeit at only 240ft.(73m). The vista, on a clear day, is one of the best on the Island; you need to walk on for a few yards, to the crest, to obtain the full benefit.

Looking south-east, St. Boniface's Down, nearly seven miles away, breaks the slope of the field, followed in the middle distance by the ridge of Bunkers Hill. Next we have St. Catherine's (once known as Chale Mountain) with, first, its Russian Monument (also known as The Hoy Monument or The Alexandrian Pillar) and further back to the right is the Salt then the Pepper Pot; or more correctly the old (unfinished) lighthouse and St. Catherine's Oratory. Modern communications paraphernalia are littered around the old lighthouse.

Almost due south and behind Whitecroft's clock tower are the hills above Ramsdown, separating the valley of the Medina from Chillerton Street; the high wood in the distance to the west of this is Tolt Copse. Next to Tolt Copse is the TV mast on the south end of Chillerton Down; the top of which stands 1299ft above sea level. Incredible as it may seem, from the top of the mast one would have a direct line of sight to a lower mast on the island of Alderney 75 miles away. In the past there has been a microwave link between the two, providing a telephone service to the Channel Islands.

The skyline to the south-west is dominated by the Chillerton and Newbarn Downs. To the west is the statuesque Great Whitcombe Manor with its backdrop of woods which, over in Hampshire, would be classified as a 'hanging', or 'hanger' wood. Farms take up the western foreground with cottages once associated with the manor; behind them to the north east is Alvington Down with its reservoirs and telephone masts. Completing the circuit you have Carisbrooke Castle and the, once Dominican Priory.

The walk continues, well defined by a wire fence, across the field to the end of the hedgerow visible ahead, at which point it joins another, shallow, sunken path, once known as Copse Lane, down to the side of Whitecroft farm. As you approach the farm, notice the base of an immense chimney stack at the gable end; it seems to have been added on, blocking off parts of the higher windows; and is now removed once more. It may have related to the brickworks for the building of the hospital?

The path now deposits you on Sandy Lane opposite the rear entrance of what was Whitecroft Hospital; originally built as an asylum for the insane, eventually becoming a hospital for those with mental illnesses.

Turn left in the lane past the farm and continue up the hill with high sandy banks on either side; as you go over the brow Paradise Farm (now just horses) can be seen in a small valley to your left. Your route carries on down the hill until a footpath (N153) is, or should be, signposted on the right almost opposite the drive into Paradise Farm.

You rise up through a now overgrown sandpit which has, in recent years, been used as an unofficial dump, although the ivy covers it well. At the top of the rise there are fields to the right, and to the left an ancient bit of woodland continues. The path twists left then right, deviating from an old sunken path that used to go down to the left. There is now farmland through the copse to the left and to the right we have the always smart and tidy, W & B football ground stemming from the unlikely pairing of Whitecroft and Barton.

At the end of this quite level section you reach more woodland and begin to descend; this I feel is one of those magical places where one must stop and soak up the ambience. Whether it be woodpeckers, bird song, distant sounds of activity or just the wind in the trees; to me there is always an 'atmosphere' here.

Storms over the last thirty years have caused a lot of damage, much of which has not been cleaned up, but it doesn't detract from the beauty. Proceeding down the now steep path you arrive at a bridge over a small stream, a Medina tributary; it seems almost sub-tropical here, all around is rather damp and there is bamboo growing on the right. The bridge is not an old one but its rustic state forgives it.

Beyond the bridge it used to be so swampy that wellington boots were essential, but a path has now been laid, follow this up past a steel gate to the front left – there used to be a split in the path here with a stile, now long gone, to the right of the gate; now just follow the path. It now passes the remains of Hill Park Dairy on the left, first a pig sty which is about all that is recognisable and then the jumble of masonry that was the cottage until it collapsed in the eighties. This was always a very dark place, built, as it was, in pit hollowed out of the hillside; although there would have been fewer trees in the past. Its terraced garden is still discernible and continues on your left. It used to be known as Park Cottage and one reference says 'Cottages'!

To the right the ground drops away to the stream with the bulk of Whitecroft Hospital behind it. It is from this perspective that one can see what a sanctuary that hospital must have been, in the heart of the countryside, for those with troubled minds especially for those who, with perhaps lesser problems, could be mended. A true asylum. One wonders if the 'quick fix' drug dependant solutions being used now, might still benefit from a spell of tranquillity 'under the clock'; to use an old Island expression.

As you break out from the woodland on the now substantial track there is a modern barn almost immediately on your left. **It is between this barn and the road a few hundred yards ahead that PLUTO crossed the path**, there is a marker here and a rare clue if you can find it.

Ahead in the distance is the fine farmhouse of Hill Farm and the view across wonderful open farmland now opening to our right includes the distant Vayres Farm tucked under the downs and, nearer, Cox's Corner; between them, hidden in a fold, is Lake Farm. Later, also to the right, you can glimpse Great Whitcombe Manor with the wall of Carisbrooke Castle behind it. More of the old hospital also comes into view; these fine old buildings have now become a gated residential development.

I cannot recommend walking the road back to Newport or on to Chillerton. In both directions there are blind corners and a distinct lack of verges. Thirty years ago car drivers might have thought that someone could be walking round that blind corner; not anymore!

If you have another hour (or so) available, you could go 100 yards toward Newport and turn left up Hill Farm Lane, this would lead you to the G10/G6 and N108 back to Newport or even further to Newport by way of the G10/N202/N205 and the N101. But most of those routes are parts of other walks on this theme.

I would recommend that you retrace your steps back to the lane, (It's just as nice a walk going the other way) and take a different route back from there.

When you get back to Sandy Lane, instead of turning left go right, down the lane; take great care to walk where you can be seen, the first bit is rather blind and there can be a surprising amount of traffic. The lane has been cut out of the sandy soil over centuries of use, forming high banks. Quite large trees cling, seemingly impossibly, to the sides.

Turn left at the next junction into Marvel Lane. Just around the corner are two modern houses. They have been built on the site of what were known as Paradise Cottages, (and before that as Blackwater Cottages); a rather poor little row with corrugated iron roofs, perhaps converted from a barn!

At the top of a short rise the lane becomes undulating and relatively straight with views over the Medina valley to St. Georges Down to the right. The lane, as lanes everywhere, is really not coping with modern traffic and in a few places it is in grave danger of literally sliding down into the field on the right. Even your average 'Chelsea Tractor' is really too heavy for it. It is only the roots of that hedge preventing a collapse. This is another hedge which is regularly beaten to death with flailing machinery rather than cut. Wouldn't it be wonderful if someone could afford to get it laid?

The left-hand hedgerow has its allied problems where the banks are in continual danger of being washed out by the sandy mud coming off the fields. The removal of hedges between fields is partly to blame, associated with the absence of the time honoured disciplines of ditching. In recent years some planting has been done in an effort to improve the situation.

However, if you can try to ignore these concerns, the walk along the lane can be very enjoyable. You eventually come to Marvel Farm, on the right, with its ancient wall which has taken the brunt of many a muddy flood. Opposite this is the track up to Great Down; the top part of which you used earlier on the outward route.

Somewhere hereabouts, no one seems to know exactly where, was a religious foundation raised, in the 12c, by Henry De Blois when he was Bishop of Winchester. It has simply been referred to as 'at Marvel' The lodge at the entrance to Marvel Farm has something of an ecclesiastical look about it but as it is 'late Victorian' there can't be any connection. It is, however, architecturally interesting with its little tower and false arches over the windows; a sort of Arts and Crafts in brickwork.

After the lodge the view to the right, across the valley, takes in first Standen House, a fine red brick Georgian building set amidst its trees, and then West Standen Farm a little further along. West Standen was once owned by the Jolliffe family noted for breeding Maj. Gen. Jack Seely's famous charger 'Warrior', when they were based at Yafford. This 'larger than life' gentleman, often referred to as 'Mad' Jack Seely, became a baron in 1926; the first Lord Mottistone.

This view is soon lost as the lane now dives to the left down to Watergate farm and the entrance to that bridleway that you used on the outward journey. From here you could simply retrace your steps back to Newport along Watergate Lane or, if you have the time and wish to make the walk a little longer, take Nunnery Lane to the left.

There seem to be various names for this lane. Perry's Guide Maps refer to it as an extension of Watergate Lane, and from the other end it seems to be called Love Lane but I believe it is Nunnery Lane in general usage and indeed that is what it is signed as.

The buildings on the right as you enter the lane, now a dwelling, were part of the Newclose establishment when it was the base for the Island's foxhounds under Mr. Harvey. After this is Brookside Dairy on the left before the dip in the lane where stand Newclose Farm Cottages by the old entrance to the farm. I suspect that there is no longer a connection.

Past here ornamental ponds can occasionally be glimpsed through the odd gap in the hedge to the right, all part of the complete reordering of the farm. Not much more can be seen until you top the rise, when Rosary Cottage and the Verbum Dei Centre can be seen on the hill ahead; once associated with the Catholic Priory.

At the top of the rise is the grand new entrance to the farm. The house is to be seen back in the valley to the right where the much smaller farm used to be. There are springs down there, I well remember that the farm had a wind driven pump, they were once quite common. (Perhaps they will return!). The grounds are maturing nicely; no longer a farm, it has the ubiquitous horses, deer (hence the high fences) and some domesticated llamas.

To the left across the valley is Whitcombe, in fact the part you can see is actually called 'Valleys'.

At the top of the next hill turn right into footpath N26, but look back and pause awhile before you do so and take in another outstanding view. The southern slope of Mountjoy hill is to the left, followed by the ridge of Staplers Heath in the distance, then you have Pan chalk pit and the bulk of St. George's Down. Shanklin Down is next with just a glimpse of Boniface. In the foreground is Great Down and above and beyond that, Stenbury Down.

On then, up to the right along the quite closed over path. On the left the high bank conceals a very old marl pit now used as a depot for the privatised Council Works. It was in this pit, on the night of the 28th May 1648, that horsemen waited to help the flight of King Charles from Carisbrooke Castle. The escape bid was thwarted by treachery, or narrow window bars, depending which source you believe!

At the top of the rise you meet the path which traverses Mountjoy hill; designated N25 to the left and N24 to the right. A whole new panorama now greets you, Alvington Down on the left, Newtown harbour almost straight ahead, followed by Parkhurst Forest just to the right. Over the stone wall is Mountjoy cemetery, 'opened' in 1854.

Turn right here, onto the N24 which climbs to the highest point of the hill. As you climb more of the north of the Island comes into view. The Parkhurst and Albany prison complex is to the right of the forest. Then Cowes, with the prominent chimneys of Kingston power station and the towers of Osborne House sit above the broad sweep of the River Medina. The south coast of the mainland is hazily in view above and behind all of this.

At the top of the hill, in the field to the right, is the ordnance survey trig. point at 275ft. (84m). As you pass the tree and the top entrance to the cemetery spare a thought for the men who spent many hours, day and night, up here during the 1939/45 war watching the skies for the enemy. Their huts were just to the right here but the area has now been absorbed back into the field. This was a Royal Observer Corps post established to give the factories of Cowes advance warning of air raids.

Walking down from this high point your view will be filled with Newport, particularly its two most grotesque structures, the corrugated metal shed at Coppins Bridge and the 'late Art Deco' blockhouse where the Bus Station was; both completely out of keeping with the rest of the town. Further down river are the vast white roofs of the industrial estate; a coat of green could have lessened their impact!

Continuing down this eastern end of Mountjoy there is scruffy 'set aside' and a horse paddock to the left with more traditional farmland on the right; watch your feet now as you get closer to the town! At the bottom of the path follow it round to the left. The large pit on that side, now populated, is the one which gives Whitepit Lane its name. Ignore the various openings on the right and you will arrive in Whitepit Lane.

Turn right down the hill, this will take you the short distance to Shide Cross where, by turning left, you will be back in St. John's Road. This leads down into Newport and back to your starting place at St. John's Church.

*Whatever you do enjoy the walk, and the stops;*

*don't forget:-*

*'A poor life this, if full of care,*

*We have no time to stand and stare'*

(William H. Davies)

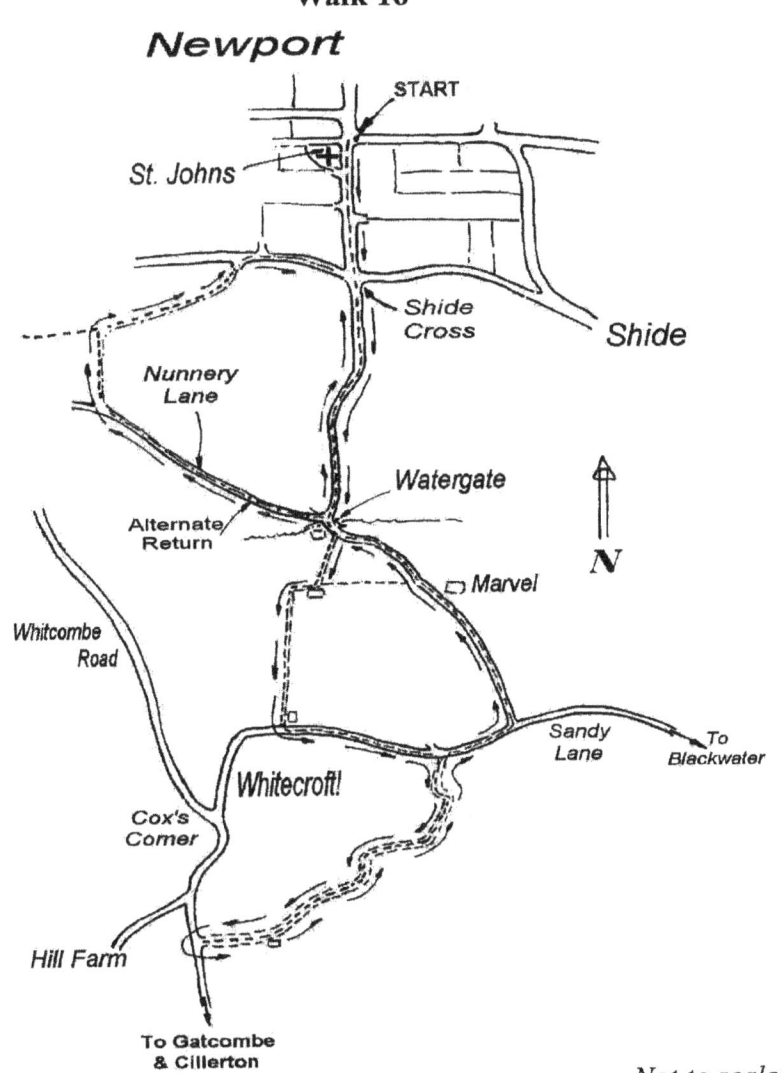

**Walk 16**

# 17. A PLUTO (SOLO) ENTRY ON THE NORTH WEST SHORE

*(A ramble south of Gurnard)*     *(Just over 3 miles)*

Start this walk from the western end of Gurnard Marsh by the bridge over the little river known as 'The Luck'. If possible, in order to take in a point of interest later, arrange to start this walk about half an hour before low tide.

Gurnard has had a variety of spellings over the centuries, starting as Gorenore in 1280; the first part said to do with marsh and mud. Was the 'nor' north, or the Old Norse 'nor' for sea inlet or the 'ora' for shore? The name 'Luck' is not, apparently, unusual and is used, in a variety of forms, referring to a pool or water held up by a dam or similar.

From the 1920s the marsh has been much built over; first with holiday huts and more recently, on both sides of the road, with more substantial buildings. It is still subject to flooding at the highest tides and some huts are raised to take this into account.

Before the first world war there was little more than a rifle range on the marsh. However, many years before that, and perhaps even in prehistory times, it is thought that there may have been something of a port here. The river is believed to have gone deeper inland and certainly the land either side of 'The Luck' protruded further into the Solent. Remains of a Roman villa, discovered in the mid 1800s, have disappeared beneath the waters and people questioned in the 19c said that people had told them, of fields two tenths of a mile out from the then shoreline. The quarrying of high quality stone from this shore over the centuries no doubt helped the inroad of the sea. (Extrapolating that level of erosion back to Phoenician times suggests that a causeway to Stone Point on the opposite shore may not have been a complete myth.)

On the western side of the Luck, walk toward the east, past Marsh Cottage, on the point. The footpath, CS17, (Part of the Island's Coastal Path) starts just up Rew Lane now; it used to go by the cottage. Marsh Cottage has been much enlarged over the last 30 years and its defences against the sea strengthened. The foundations of the Roman villa lie off the point; part of it in what was once the cottage garden. There is known to have been a small, probably earthen, fort hereabouts in the 1630s. The path rises along the top of the cliff, meeting the old (collapsed) footpath at the top. Being the narrowest part of the Solent many services, including power, telephone, gas and water come ashore in the Gurnard area. Charts of the area proclaim that anchoring is forbidden; due to the risk of damaging these connections.

The cliffs here change shape all the time and the path has to be moved or bridged occasionally. Unfortunately, these paths, along the top of similarly composed Island cliffs, tend to have a self-destruct element. The turf being worn away allows water ingress and, in dry times, cracking, both of which, with the pounding of feet, tend to cause the cliff top to break away at the path. Not a safety matter, on most of this walk, but be particularly aware between Freshwater and Chale. There are other, more isolated, examples on other stretches of coastline.

Another stile between abandoned huts leads to a more open section of the path with a field to the left containing, on the far side, more huts of varying ages and usability. The cliffs to the right are heavily grown over, which slows, but doesn't prevent, the slipping of the clays. Across the Solent can be seen the wooded coast of Hampshire and the vertical complex of Fawley refinery. Due to the fields not really being farmed, and therefore the lack of pesticides etc., these first two sections of the path have a wealth of wild flowers and, in September, are good for blackberries. Sadly, Japanese Knotweed seems comfortable here as well.

Occasionally noticed on this stretch, a partially concealed path leads down to one or two huts nearer the beach; usually only glimpsed in winter, unless you make a foray. The coast here is very rocky and used to be good for prawning; ill-advised now due to the presence of heavy metals. The sea views here are rarely boring, there is usually something going on; ships often anchor in Thorness Bay, ahead. The concrete slabs you pass over on the path are the remains of WW II look-outs.

The route now needs no directions, just follow the cliff top path. At the end of the field you leave the holiday huts behind and start passing ordinary farming land. The cliffs to the right now become more heavily wooded with a huge variety of saplings, all helping to stabilise, at least the surface of the slopes. Good to find sloes here.

A small copse of stunted oaks is next then more open farmland with Hornhill Copse over the fields to the left. Occasionally it is tempting to try one of the apparent ways down to the 'beach' but beware, unless you have detailed local knowledge, don't; Blue Slipper and mud slides may be waiting to claim you!

After a longer, more enclosed portion of the path, with some quite mature trees, you emerge again by the side of fields and in a few yards, the whole panorama of Thorness Bay opens up.

The view straight ahead is Brighstone Down with Rowridge TV mast and to the right of that is Chessell Down. The high ground further right still, above Thorness Holiday Park is the northern slope of the Main Bench by the Tennyson monument. Still to the right after that, somewhat nearer, is Hamstead and further out Totland and Headon Warren. In good visibility, further round again, the hills of the 'Isle' of Purbeck can be seen across Bournemouth Bay.

The path now follows the edge of the cliff around a clump of bushes, after which more of the cliff can be seen, and the height and breadth of the unstable mass can be better appreciated. As you come to the highest point of the path you will start to see some bits and pieces of masonry in the cliff fall and at the top, partially concealed by bushes, are the ruins of a WWII searchlight position. The generator shelters are still there, those to the right just about hanging on to the cliff edge. To the left, inland, past some more holiday huts, can be seen the blockhouse and magazine, complete with blast walls, of the Anti-Aircraft Battery which was also located here on this very commanding position.

The path turns down, to the right here, between the generator shelters, with the more noticeable ones to the left. As you clear the bushes the path continues downward with more of those private hideaways to the left, even an old railway carriage. (How did they get that there without the mobile cranes that we now take for granted?) The view widens out to the left now, with Chillerton TV mast coming into sight. At about this point you cross the boundary between Cowes and Calbourne and the path becomes the CB24.

As you near the shore, the path goes through a very disturbed area of the cliff; it will soon be necessary to realign the path, further in.

There is no mention of Swinburne having walked these particular cliffs but 'twas just the same in his day, as his lines explain: -

*Till the slow sea rise and the sheer cliff crumble,*

*Till terrace and meadow the deep gulfs drink,*

*Till the strength of the waves of the high tides humble*

*The fields that lessen, the rocks that shrink,-*

From 'A Forsaken Garden'

It is almost as though he was prophesising global warming!

Back to walking. After this difficult section the path levels out above a low section of cliff and crosses a small footbridge. The path continues to skirt the field and comes to a further little footbridge.

From here, at low tide, down on the beach to the right will be seen a collection of some nineteen vertical stumps, the remains of concrete posts. This was the location of the manifold which 'collected' the many, small diameter 3 inch diameter cross Solent pipes, code named SOLO, which brought the fuel to be pumped through PLUTO from the mainland. From this collecting manifold the fuel went into a nearby pumping station to be pumped across the Island to Hungerberry Copse above Shanklin.

At very low tides the remains of some 12 or 13 of these pipes can be seen snaking out across Quarry Ledge, having come from Stone Point on the distant Hampshire shore. Also visible at very low tides are remains of the groynes made up of concrete blocks put in by the sappers to protect the pipes.

After a while the Coastal Path (now CB24) continues along the cliff top, but decreasing in height it goes down onto the beach. However, a permissive path allows you, by way of a stile, to continue inside the field parallel to the beach, in front of a crescent of holiday huts. About halfway across the field is a fence and a further stile, after which a footpath leads off, at right angles, to the left across the field (known as Sea Close) towards the centre of the huts; this is a proper footpath (CB1) stemming from the Coastal Path. (The permissive path, and Coastal Path on the beach, continue ahead)

**Somewhere about half-way between the first stile into the field and the stile in the centre of it PLUTO must have crossed the path**; (albeit at a very acute angle) there are no markers. (I have been unable to establish where the pump housing was located, but it is more likely to have been in this field, well above high water, than on the exposed beach!)

**After you turn left onto CB1, after that second stile PLUTO would have again crossed your path on its way diagonally across the field.** As there was no boundary to the path there would have been no markers. After the field, the path crosses the access track in front of the huts and goes straight between two huts, through a small coppice, over a 4th stile into a field, and continues with the hedge on the left. This section may be muddy.

Continue up alongside the hedge; this field is known as Thistle Hill. The farm to your right front is Whippance Farm and your path with its line of boundary Oaks is going toward Sticlett Farm which can be glimpsed to your left front. This path may be that used by Coastguards who were based at Sticlett between 1840 and 1870 and would have kept a boat near the beach; it depends where their cottages actually were. In 1862 there was also a path to the beach, slightly further north, from the older part of Sticlett,

The gantry away to the left, as you reach the 2nd field, is where one of the Cross-Solent electricity cables has its transition from underground to overhead power lines.

Someone has tucked a summerhouse into the oak coppice on the left a little further on. I wonder if those Coastguards had a lookout post here; it gives an excellent view of the bay. A stile takes you into another field, called High Gate, across which another stile can be seen in the top left corner

The path from here is to the left through the copse. (Not through the gate onto the track, that is not a FP). It's worth a look back from the stile as the views will be lost for some time now; a seat has been thoughtfully placed just inside the copse.

Sticlett does not appear to have been a Domesday Book manor but the name, to do with the stream emerging below Whippance Farm, is documented as having been associated with the area since Saxon times.

The path through the trees emerges between the gardens of Sticlett cottages. Rose Cottage on the left may be associated with that summerhouse passed earlier. Just past the house on the right the path is crossed by a farm track. The path you have been on, the CB1, turns right here, to Hillis Farm. To the left the track goes through the farmyard and to the older Sticlett which is about 250 yards further on.

However, your route now becomes the CB23, which starts straight across the track where a stile takes you into more woodland but in a few yards, by means of another two gates, soon crosses another track. The path continues through woodland, initially running alongside the cattle sheds of this modernised dairy farm, well known on the Island for its milk.

The woodland is mostly oak interspersed with the occasional ash tree. At the end of the wood a stile takes you onto a fenced off path. At this point you recross the boundary and your path becomes the CS3. From the stile you have a wide view of Parkhurst Forest and nearby countryside.

Just peeping above the forest, ten and a half miles away, is the top of Shanklin Down. The houses immediately to the right are those clustered around Hillis Corner. The farm in the fields to the right front is Skinners Farm; behind and above that, are the red roofs of Pallance Farm.

As you go down the path the field to your right is known as North Close. Near the road in this field there was a house, where the bungalow now stands, known in the 1800s as the Manor; not a true manor, previously it had been called White Hall.

The path terminates at a stile by a capped off well, after which you are in Rew Street, believed in the past to have been part of the mythical 'Tin Road' (maybe not so mythical!). From the stile, slightly to the left, a tower appears on the skyline; this is The Cowes Radar Factory's contribution to the scenery. The tower was used for testing aerials and it would be nice to see it remain as a symbol of the employment given to Island engineers in the fifty years since we stopped building large ships and aircraft.

Turn left up the road. Immediately on your left is a footpath (CS20) to Pallance Road (or Tinkers Lane as it was once called). Take this path if you wish to return directly to Cowes; perhaps with some refreshment at The Travellers Joy. However, if you wish to return to your start point continue up the road. The road is not quite as straight now as it appears on the map but it is doubtless an ancient way from, in centuries past, one of the Island's main points of entry at Gurnard. It may be Roman but could be earlier!

There are a number of long established farmsteads along this road to the coast, but as will be seen there has been a lot of ribbon development between them during the 20c.

Dukes Farm is first on the left; its farmhouse is next door with its small barn attached to the side. Shortly after this Harts Farm comes up on the right with its barn end of unusual bricks facing the road. Next on that side is Bridle Cottage, closely followed by Harness Lane!

A little further, also on the right, is the somewhat more substantial old farm known as Rew Street Farm. In the not too distant past, the road appeared to go through the farmyard here with its duck pond (now ornamental) and cart shed on the left and the main buildings to the right. The farm has a fine range of stone buildings and a tall, three story high, Georgian farmhouse. The original farmhouse lying alongside it, end on to the road, was of cruck frame construction and the two timbers forming the cruck remain in situ but are now buried beneath the new stone facing of the gable end.

The cruck frame construction at Rew farm is believed to have been the only example on the Island; certainly the only one left.

The tastefully converted cart shed on the opposite side of the road is contemporary with the old house and I am fairly certain that I remember seeing a cruck frame in the southern gable. The portals of the cart shed bays with their naturally formed timber knees remain, facing the road, forming a sort of extended porch for the 'new' dwelling.

The last farm on the right is Basketts Farm, another old farmstead and in rather an exposed position; the roadside barn having, noticeably, taken the brunt of a few centuries of south-westerly gales.

After Basketts there is a short stretch of undeveloped countryside with rising ground to the left and the valley of 'The Luck' to the right. It is believed that boats could come a considerable way up 'The Luck' in the past but no dates seem to be available for when this was possible. In the valley is another of those gantries where cross Solent power cables emerge and become 'airborne'.

Part of the slopes on the other side of the valley were the grounds of a large house called The Dell which was turned into a holiday camp called Gurnard Pines in 1939; it has been much upgraded and is now quite a smart holiday and leisure location. During the early months of 1944 the original house and grounds became a military camp supporting 'J' Force; one of the 'D' Day formations. The original house (it was only 58 years old!) was demolished after the war.

Gurnard village centre is on the ridge across the valley with the small, refurbished spire of All Saints church prominent above the houses.

The road now comes into an area known, for some reason, as Horn Hill and starts to drop down to the coast. 'Hillside', a much-improved old house comes up on the right followed by a nicely situated bungalow. The last farm, Cliff Farm, is on the left at the bottom of the hill, as the road bends to the right, after which you are back where you started at the rebuilt, but still very narrow, bridge over The Luck. Buses stop here but only about four times a day!

It is possible to walk round the coast to Cowes by going over the bridge and taking CS35 to the left along the sea wall and beach to Gurnard proper; but at high tide you would have to return to the road at the end of the marsh.

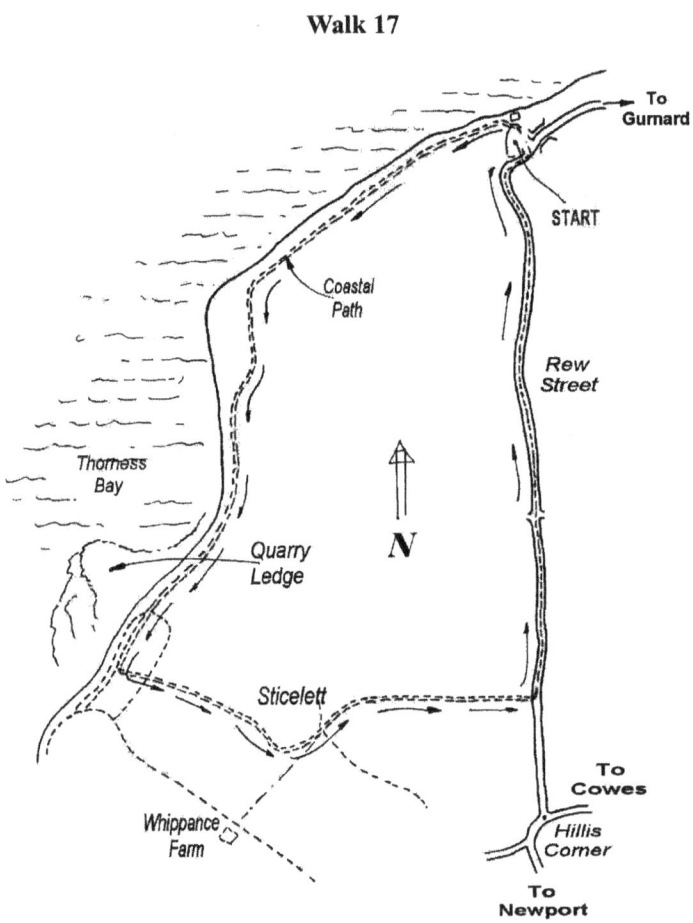

*Not to scale*

## 18. AN EVEN SHORTER SHANKLIN CIRCULAR

*(For the sake of completeness)    (About ½ a mile.)*

Starting from the car park near Shanklin Old Village, the one with the War Memorial tucked in the corner, by the road. Go down the hill on the main Shanklin to Ventnor road; this is pure 'Grockleville' with little more to be said, although there are some gems amongst it all.

Go down beside the Crab Inn and past the lane down to the chine on the left; note the fountain here, commemorating the visit of the American poet Longfellow with his inscription and the British and American flags.

The 'Village Inn' is on the right and behind that is a small mews area of craft shops. Stay on the main road around the bend and continue out of the village on the right hand pavement.

The area is notable for its thatched roofs, some old, some new, almost all now are thatched with reed rather than the straw that would have been traditional here. (And the reed probably having come from abroad rather than Norfolk!) Many of the older houses also have very fine woodwork, especially around the eaves, which has been surprisingly well maintained. The trees and other plant life are particularly vigorous here and was much commented upon by Keats and other men of letters during their sojourns in Shanklin.

As you leave the Old Village reflect on the fact that this is a relatively new road and was only made the road to the church of St. Blasius and on to Bonchurch and Ventnor in 1826. Before that you would have had to go through Chine Hollow and by way of Luccombe or away to the right via the Manor Road and by the Manor Farm.

Continue past Rectory Road on the right and you will shortly come to an area of open grassland also on the right; this is known as 'Big Mead'. **Somewhere between the beginning of the grass and the turning into the road on the left PLUTO crossed the path (in this case, road).** No markers are to be found and possibly there never were any.

There is a suspicious looking depression running through 'Big Mead' at this point but I don't think that its origins were to do with PLUTO, it may, however, have been made use of.

The pipe here would have been the gravity feed from the great storage tank, in the trees some half a mile away, to the pumping stations down on the level of the promenade. After crossing the road, the pipe presumably went straight down into the chine and commenced its tortuous route through it.

Cross the road at this point to turn into Priory Road on the left. Note the memorial plaque in the wall on the opposite corner. This is to the men of 46 Commando Royal Marines who used the nearby Upper Chine Girls School as their HQ when in training for 'D' Day. (Much of the school was demolished in recent years and the site developed!). There is a tale relating to the signs in the dorms 'Please ring this bell if you require a mistress in the night' - much pressed by the marines we suspect?

As you enter this road you actually go over a bridge over the chine. A look over the hedge to the left will reveal the downward path the pipe must have taken into the chine. The house to the left has made the most of the ravine by turning it into a sunken garden.

There used to be a theatre to the right, known as the Margaret Passmore Theatre, and later called The Portico. It was demolished circa 2010 and is, at the time of writing, just wilderness. Just past it is a path to the right which leads up to Shanklin's old Church of St. Blasius, avoiding the main road.

However, your walk continues round to the left, entering Popham Road which runs parallel to the chine. Priory Manor Hotel to the left is basically Victorian, although currently derelict, otherwise the houses in the first part of the road are more modern with a particularly interesting use of red brick alongside the road on the right.

At the bottom of this road bear to the left leaving Rylestone Gardens on the right; this will take you down into Chine Hollow. The term 'Hollow' is not much used on the Wight but it is quite common in Dorset.

(This has been closed for some time, at the time of writing; it is possible to go beside it within the gardens and take steps down to bypass the closure.) This lane was still used by motors until 1930. The near vertical sides are covered by ferns and tongues, all very verdant even in the driest weather. The roots of large trees high above are miraculously hanging onto the sandstone cliffs and at the bottom of it all there is a mini ford with the stream running over a paved base.

**It must have been about here that PLUTO crossed the path for the last time before continuing down the chine.** No markers here, it's doubtful if it the pipeline was even buried at this stage, as there being plenty of natural camouflage and the area would have been completely under military control.

You may be able to walk up the chine a little way to the left and this leads to the King Harry bar. However, although there is nothing to be seen of PLUTO in that direction, an appreciation of the route it must have taken, through that part of the chine, can be seen. There is a gate on the left, possibly disused, a little further on, which also leads to King Harry's bar (where a drink and food may be found). Crossing the ford and proceeding instead to the top of the Hollow brings you back to the centre of the Old Village alongside 'Longfellows' fountain, outside the Crab Inn. He came here in 1868. On this short walk you will not have justified his opening line but you may be ready for his second: -

'Traveller stay thy weary feet;

Drink of this fountain pure and sweet'

I don't think I would have trusted it in 1868, without knowing what was 'upstream'. There is far better ale to be had in King Harry's Bar. At the fountain turn right to return to your starting place.

**Walk 18**

*Shanklin*

*Not to scale*

# 19. UP HILL AND DOWN DALE IN TOWN.
### *(A bit of countryside but mainly developed, about 2 miles)*

Off the Avenue in Shanklin there is a road signed as Chatsworth Avenue leading to Orchard Road. This is rather puzzling as the road markings later suggest that it is actually the other way around. However, never mind that. There is an interesting little walk that can be started from here, and there is some 'on road' parking available nearby.

**I believe the PLUTO feeder pipeline to Sandown Fort must have crossed The Avenue somewhere near this location, but can find no indications of its passing, so please have a diligent look as you progress.** Start by taking Orchard Road, down in the dip then up again. All of the housing hereabouts except those actually on The Avenue have, of course, been built since the war but there were previous tracks and paths on and over Sibden Hill, which is the high land to your left as you proceed. Continue just a little way up Orchard Road but bear left, as Orchard Road bends to the right, onto an unnamed, but paved lane, there is a FP99 sign partly hidden by trees on the left. There are some concrete posts hereabouts, annoyingly similar to PLUTO pipeline markers, so do not be deceived, but there again you may just happen upon a genuine one, so good luck.

After about sixty-five yards, as the paved lane is curving to the right, leave the lane to the left up an obvious path through the trees; this is not signed but is still the SS99. Keep straight ahead up through the trees for about fifty-five yards and where the path forks, keep to the left and you shortly reach a east-west path.

However, before turning right (east) (mark this point) it is worth having a look around Sibden Hill, the summit of which is now above and behind you to the left. There are a number of rough paths around the hill here, which is registered as a Site of Importance for Nature Conservation (SINC). The paths are not marked but the hill is too small to get lost on. At the highest point there used to be a Flag Pole (the supports of which remain) used as an Intersect Station for the re-triangulation of England in 1936. Before that in 1931 the hill was also one of the stations used to record the climatic details in a drive to promote Shanklin's wonderful climate and especially the sunshine figures. (In 1939 the roof of the Town Hall took over). For many, the flora and fauna of the hill will be the more interesting, but for some the search will be for certain concrete posts.

For a piece of poetry to accompany this walk, I was hoping to quote something allied to John Keats' stay at Shanklin in the late summer of 1819, but he seems to have had women on his mind at the time rather than the countryside. However, at Newport in 1817 he started his greatest work 'Endymion', from which I would like to quote just two lines from his four thousand plus; the first and last: - (The first often being quoted with reference to The Island of Wight.)

'A thing of beauty is a joy forever'

and

'Home through the gloomy wood in wonderment'

After exploring the hill, return to that 'marked' position and regain the SS99 FP, go downhill in the wood margin with the field nearby on the left, beware it is steep in places and can be slippery when wet. **If the PLUTO pipeline to Sandown went east of Sibden Hill and west of Batts Copse, it must have crossed this path!**

The path turns left at the end of the field and runs with allotments on the right before again turning right and going downhill, after which it goes left for a little way with an un-named stream on the right. This path ends by going down a few steps to crossroads of paths with a bridge over the stream on the path to the right. Our path will be uphill to the left, the SS14. Before going up the path to the left, a look over the bridge may be of interest, there are four pipes crossing the stream and it is tempting to think one of them could be for PLUTO, one is certainly about the right diameter but none is encased in concrete like other PLUTO crossings and it is difficult to believe that the pipeline would be buried to the same depth as these pipes, especially in the northerly direction. (but just maybe !!)

Proceeding up the path to the left, going north(ish) Batt's Copse is on the right and a relatively recent housing estate on the left. The path follows the course of an old track, half of which has now been left as a ditch but the bank on the left shows the width of the original sunken track. The path comes out where the end of the pre-war housing meets the modern extension of Carter Avenue. Both of the paths we have followed so far are ancient pathways, and were there long before any of the housing estates in this area. The SS99 and the SS14 were the Urban District Boundary and the SS14 was also an old track over what was known at Hide Hill. The SS14 crosses Carter Avenue to the opposite side of the road, on almost its original alignment, but now bordered by high fences and holds no interest whatsoever, quite de-naturalised, and after quite a short distance it emerges onto Hyde Road, a much older thoroughfare.

Turn left onto Hyde Road and then almost immediately right across a bridge which used to carry local, mainly agricultural traffic over what was the Shanklin to Ventnor railway. The path is now signed Footpath to America Woods and is actually the SS92. **At the far end of the right hand brick parapet there are the two concrete post of a PLUTO pipeline marker showing,** I believe that the pipeline came over the bridge, to cross the railway, which was then very much in use, and left the bridge where shown by the markers. The presence of another single post, part of a pipeline marker, on the same side, about four and a half yards further on is something of a puzzle.

Continue down the path to a meeting of the ways, a crossroad of paths. Ignore the various entrances to the blight of caravan positions and take the Byway SS87 to the left. This is another old track which eventually runs out next to Cliff Bridge on the main Whiteley Bank - Shanklin road. But we are not going that far. This whole area was part of the original 'hide' of Upper Hyde Farm, the spelling seems to vary a little. This is an ancient settlement and the 'hide' being considered the amount of land required to support a family. (roughly 120 acres but it varied with counties and land quality. One wonders if the size of the family was ever considered!). The spelling at Domesday was Hid or Hida, how the 'y' crept in is anyone's guess, a cartographic error or an academic edict!! Together with Hide Hill, earlier, we seem to be stuck with both spellings.

Proceed along this track, there is a very deep ravine to the right which at first appears to be another sunken track-way but turns out to be a watercourse which at some time long past carried far more water than it does today After about a quarter of a mile it eventually terminates in a spring. Carry on along this track until you reach an obvious and significant gap on the left which, at the time of writing, has a white wicket fence either side of a gap. This used to be a level crossing into the fields of Sibden Hill, over the railway which has been running in parallel on the left. Go onto the old railway at this point and turn back left onto it.

The last train ran here in April 1966, the line having opened in 1866. Continue for about 150 yards but then take a gravelled path to the right, it's not signed but there is a blue sign on the left, opposite the gravel path, about 'path sharing'. This path will take you out to where Carter Avenue meets Blythe Way. Now turn right and go up and across Blythe Way about 100 yards and turn left into a cul-de-sac (also marked on Google as Blythe Way for some reason). At the end of the cul-de-sac enter the field and go up diagonally to the right. This path is not formally marked but appears to be open access.

This takes you up to the northern flank of Sibden Hill with the opportunity to look back from what is a wonderful view-point for east Wight. Fortunately, very few of the caravan parks are in view. The downs above Brading are to the left, Bembridge Down and Culver Cliff to the centre and the hills to the east of Portsmouth between the two. To the centre right Shanklin is tucked below you and the coast with the ubiquitous few ships at anchor in the bay. Beyond, on a good day it is possible to pick out the Sussex coast, certainly to Selsey Bill.

There are a few small/short paths to explore on top of this part of the hill, but to continue the walk, enter the wooded area and take the path to the left going slightly downhill, keeping the field boundary a little way to your left-hand side. Ignore paths to the left, back into the field, and small paths up into the trees to the right, until you reach a point where a stile can be seen going into the field, about fifteen yards to the left. Here, to the right, is the path you came out of earlier, before going down through the wood. (The SS99) (Although unmarked at this point). Take this path to the right to retrace your steps back to the starting point.

**Walk 19**

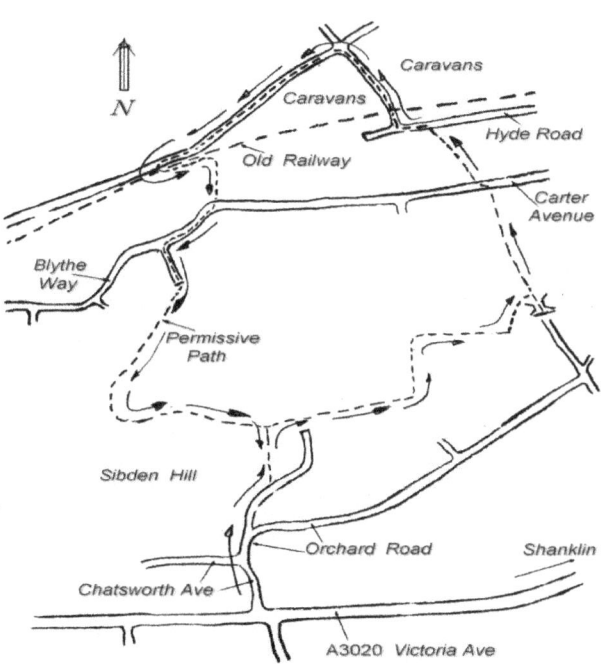

# 20. WESTERN LAKE AND THREE MANORS

*(Another mixture of field and 'development' about 2½ miles)*

I have to admit, without, I hope, offending the residents of Lake, this walk is rather lacking in scenic beauty. Initially the walk is short in PLUTO interest but it has, later, two of the very rare artefacts associated with the Shanklin to Sandown section of the pipeline.

Start the walk west of the roundabout at the Lake business park and just west of Scotchell's Brooke. This is at the junction of the lane into Sandown Airfield or Lea Airfield, as it is sometimes called. The lane is actually named Scotchells Brook Lane and it is also classified as a bridleway to Borthwood Copse and Queens Bower (SS57). One of the houses on the right stands on the site of Sandown and Shanklin's Isolation Hospital, which was in use from 1899 to 1933.

A little over 100 yards up the lane a footpath (SS28) takes off to the right along a boundary, a rather uninviting path to start with. Some development seems to be taking place to the left of the path and across to the left behind that is the rebuild of the burnt down Aviator Restaurant, will it reopen? After a short distance the path opens out with a towering macrocarpa tree to the right and you then enter a gravelled area at the back of some airfield buildings, continue across the gravel to a track on the other side which shortly turns to the left. Carry on up this track towards the 'mini' aircraft hangers and you arrive at the junction of your path, the SS28 and the NC22, and of course the boundary between the parish of Sandown and Shanklin with that of Newchurch. Take the NC22 to the right.

This airfield has had a very mixed history, silly as it may seem to say. It started in 1929 someway to the south at the fields of Apse Manor Farm towards Rill. In 1934 it relocated to fields between Landguard Manor and Ninham but this was very brief and in 1935 it 'reached' its present position on the fields of Lea Farm. The airfield was closed down in 1939 but reopened after the war with some scheduled air services which had some initial success but were never ultimately viable. However, it has thrived as a private airfield and has interesting activities taking place around its periphery, not least of which is the manufacture and refurbishment of parts, including airframes for veteran WW2 aircraft, especially Spitfires.

Your path now meanders along the eastern boundary of the airfield with Brading Down as the skyline ahead, and the buildings housing those peripheral activities are across the airstrip to your left front. This part of the path ends at a crossroad of paths, the actual signpost is some 30 yards to the left at the edge of the airstrip, but for this walk go right here through a field gate. This is the SS25 (the sign which you have by-passed says it is to Merrie Gardens ¼ Mile) Keep the hedge to your left along the boundary and at the end of the hedgerow you come into a very flat, potentially marshy area, and after crossing something of a causeway you arrive at a stile over which is the very marshy start to the southern part of Black Pan Common. How wet your feet get will depend on the time of year! In a short distance a footbridge leads onto higher ground.

There are some informal paths here but keep to the main path which climbs very gradually to the right, keeping to the right on this path brings you to a small clearing in the bushes and shortly after this, some 30 yards, you reach a fork, take the left fork. The right fork leads to Merrie Gardens, of which more later.

This immediate area is the location of Black Pan Farm, an ancient Domesday Manor, spelt then as 'Bochepone' there are some ruins, rarely seen due to being almost totally overgrown, but it was farmed well into the last century. From this point it is worth examining the hedgerows for any sign of PLUTO pipeline markers although it is believed to have passed a few hundred yards further east; nothing has been discovered so far on this route. **Continuing now in an easterly direction you will shortly see the playing fields of what is now Broadlea Primary School on the right, it is known that the PLUTO pipeline crossed these fields which then, of course, were agricultural.** Next on the left, the path is bordered by some modern fencing which is the boundary of another more specialized school which burnt down in 2015. The site has now been deemed suitable for housing.

As you approach the end of this path take note of the Nissan hut in the corner of the school/ council property to the right, surly not a WW2 leftover on the line of the pipeline? Houses now come into sight and on reaching the end of the path continue onto the left-hand pavement, this small area of post war houses has been given the name of Berry Hill. Continue alongside the cemetery until opposite the turning into Manor Road. **Somewhere along the route just taken, and probably in the last 200 yards, the PLUTO pipeline must have crossed the path, possibly skirting the cemetery.**

Cross the road and proceed south along Manor Road, at the end of which you meet the main Sandown to Newport road (A3056), turn right and go up the road to the pedestrian crossing, passing Sunnyhill Close on the way.

Cross the main road at the crossing and join the footpath SS96 opposite. However only go up the path some 10 yards before turning sharp right to take the path which is alongside, but above, the main road going west. A somewhat better view up here, the downs away to the right the airfield down in the dip to the right front and the trees around Cheverton Shute ahead.

**Continue down the path for approximately 250 yards and in the hedge on the right, opposite the first part of the white house across the road you will find the two concrete posts of a PLUTO pipeline marker, the wooden crosspieces are missing and one post is ivy covered but this is one of only three positive marker positions found on the route of the Shanklin (Hungerberry Copse) to Sandown (Zoo) pipeline.** A further 80 yards will bring you to the roundabout on the main road.

This area, known as Merry Gardens takes its name, from the 'merry' trees which grew across this area. In some older dictionaries 'merry' will be defined as 'gean', the name of the wild European cherry. But 'merry' (or merrie) actually derives from the old French for cherry, 'merise'. In some wild cherries the fruit can be quite dark, which probably gave rise to one local explanation that the 'merrie' trees were the black cherry.

Turn left at the roundabout onto Whitecross Lane, cross to the right-hand side as soon as possible and keep going to the top of the rise, one could hardly call it a hill. At the top where the road bears down to the left you will find a junction of footpaths on the right, do not take either yet but this is where the PLUTO pipeline crossed Whitecross Lane, it has been much widened since 1944 and has been developed on the east side so any markers will be long gone. Continue now down the road and after a few hundred yards after a dip in the road fork right up a bridleway, the SS89 signed to Ninham Branstone and Godshill, (Some maps show this as Whitecross Lane) this goes through the old estate of Landguard Manor. Go past Lilac Place of 1878 and the track bears right, passing the back of the old manor grounds on the left, glimpses of the manor can be seen. This was also the location of the estate saw pit before the days of machinery. You can now enjoy some woodland scenery, go down and over the bridge and as you are approaching the believed route of the pipeline again it's worth looking for any signs of markers. A quite massive beech tree, a little beyond the bridge on the left is worthy of admiration, it really is a giant.

Carry on up the track, there are caravan parks on both sides now which will have probably done away with any indications which may have been there in more agricultural days. However, go up the track to the crossroads, straight across is to Ninham but you need to turn right here onto the Public Byway SS18 which says it is to Whitecross Lane! (No, that would mean taking to the footpath later) The SS18 which comes from the left, as you join it, is the most likely route of the pipeline either under the track itself or either side. The side to the right is unlikely due to the 'pre-war' trees, either roots or getting a straight line would have been a problem (which is why the pipeline didn't go through Parkhurst Forest).

As you progress north along the track the woodland to the right is quite a picture, helped by the variety of trees and the various levels. In a few hundred yards just past where there is a gate across the track, you will arrive at a little bridge, look over the right-hand parapet and you will see a slab of concrete bridging the stream parallel to the bridge you are on. This, it is believed, encased the PLUTO pipeline, this knowledge is thanks to Tim Wander who has established that similar situations are to be found in the Romney Marsh area where PLUTO pipelines heading for the Dungeness pumping stations also had to cross streams.

**A short distance, some 10 yards further on from the bridge, sometimes just partly visible, is a PLUTO marker concrete post buried at the side of the track. This undoubtedly came from the hedgerow to the right where it is known that the pipeline left the track to cross the field to the right.** Turn up steeply to the right here onto the SS21 to cross the field, the path follows the line of the pipeline up to Whitecross Lane. This is now a very rough piece of ground, obviously destined to become something even worse.

**At the end of the path you again reach the point where PLUTO crossed Whitecross Lane but to avoid retracing your steps and to get just a little more exercise take the other path (SS22) back across the field to the north west.** This borders the rapidly expanding 'Business Park and takes you back to the Public Byway SS18 and on emerging onto the Byway the scenery improves with the mass of Shanklin and Wroxall down to the left around to the adjacent Stenbury Down. This Byway on old maps was referred to as The Old Highway, presumably between the Apse Heath area and Shanklin, it is certainly well cobbled in places, a roundel nailed to a tree now claims that it is part of the Chalk Ridge Trail. Turn right on the Byway and carry on until you meet the main road near to where the walk started. At the junction is Scotchells Lodge, old maps reference it as 'Lodge' and I suspect it was originally built as a lodge to Landguard Manor. Also in this location, in the 1080s, was the Domesday manor of Scaldeford named for the then shallow ford over Scotchells Brook.

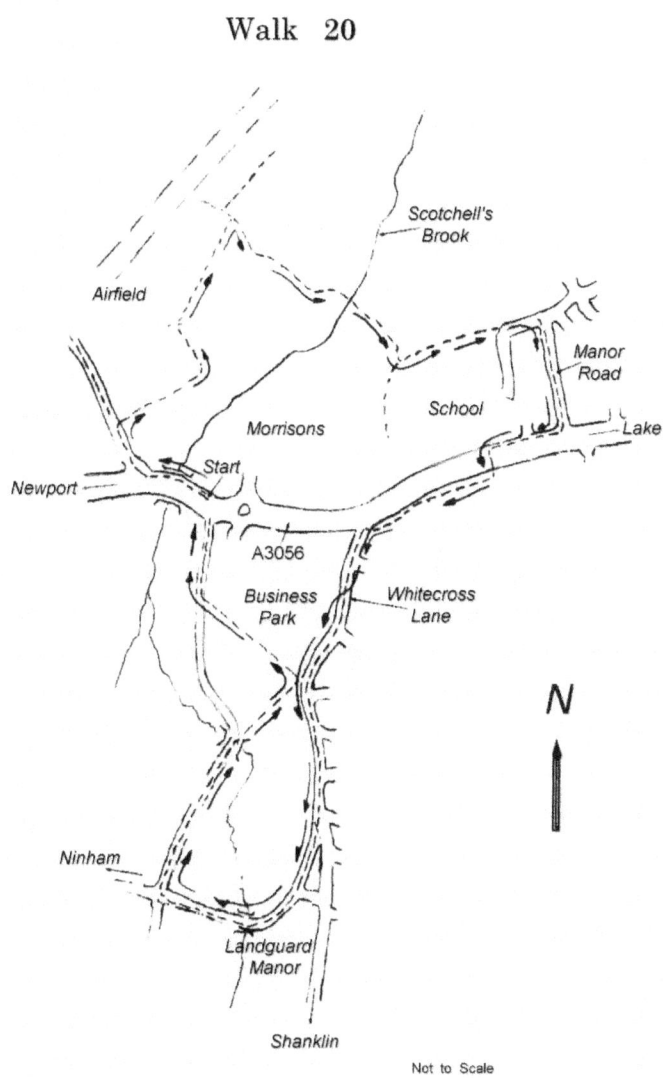

# THE ROAD CROSSINGS

## *(Some Notes just to tidy up)*

This bit is by no means intended as a walk, it is just that the reader may well ask 'what about some of the roads?'

Where there are, or were, when I last looked, definite remains of markers I have managed to include all except one in the walks. With that one exception those roads which have been missed have no markers and, as it happens, those are mostly the ones too hazardous to include in a sensible walk.

However, as the reader will have gathered, there are a few notable roads missing. I will therefore run through all of the road crossings to say what I think the situation to be.

* Indicates that these crossings do have PLUTO markers or remains of them to be seen.

The Cowes to Porchfield road * is included in walk N° 17

The Newport to Yarmouth road * is included in walk N° 10

The Newport to Calbourne road is not included. Although the section of road which would have been crossed is safe to walk along, the sections on either side of it are not.

**PLUTO would have crossed somewhere between the entrance to Alvington Manor and the entrance to Reads Farm.** The road has been widened since 1944 and no markers remain.

The Newport to Shorwell road * is included in walk N° 14.

The Newport to Chillerton road * This is the exception to the rule. I have not included it in a walk because the vicinity of the crossing is just too dangerous. **PLUTO crosses the road just to the Cox's Corner side of the Hill Farm junction.** There is a part of one marker but if you do go looking for it please do so at four thirty am. on a nice sunny morning in the summer when there is no traffic!

<u>The Newport to Godshill road</u> * is <u>mentioned</u> in walk N° 2. This road has also been widened since 1944 but part of a marker does exist, on one side.

<u>Merstone Lane</u> * is included in walk N° 15.

<u>Lessland, or Bathingbourne Lane</u> * is included in walk N° 11.

<u>Canteen Road</u>, from Whitely Bank to Apse Heath is not included. I do not know its history but I suspect that the section which was crossed has been widened. Again, until very recently (late 2006) it was not a sensible walk either side of the section concerned. Even now the Whitely Bank end is hazardous.

However, **PLUTO would have crossed the road between the last houses of Whitely Bank and the first of the houses around Pear Tree Farm.** I can find no remains of markers here.

<u>The Whitely Bank to Shanklin road</u>. This crossing is more of a puzzle. It is definitely not a road to walk along so I have not attempted to include it in any of the walks, apart from crossing it.

**PLUTO crossed it at a very acute angle between the property at the high point in the road (known as 'Island View') and the next track to the right, going towards Shanklin.** I can find no sign of any markers. I would be surprised if this road has been widened as it is relatively new, only having opened in 1885 and it is not particularly wide. However, the hedgerows do not seem so old, which could explain the absence of any remains.

<u>The Shanklin to Ventnor road</u> is included in walk N°18. (This also includes Chine Hollow which is no longer a motor road)

# 21. THE SHANKLIN PUMPING STATIONS -

## *A SHORT STROLL ALONG THE SEA FRONT*

The purpose of this short tour is to introduce the walker to the easily accessible Shanklin PLUTO pump site.

A good place to start is to stand on the remains of the pier located just before the clock tower. Ample parking is available roadside or across the road on the site of the Royal Spa hotel, by the cliff lift. **A plaque on the pier head commemorates the PLUTO project. Standing here you are literally standing on PLUTO and where it crossed under the road in front of you.**

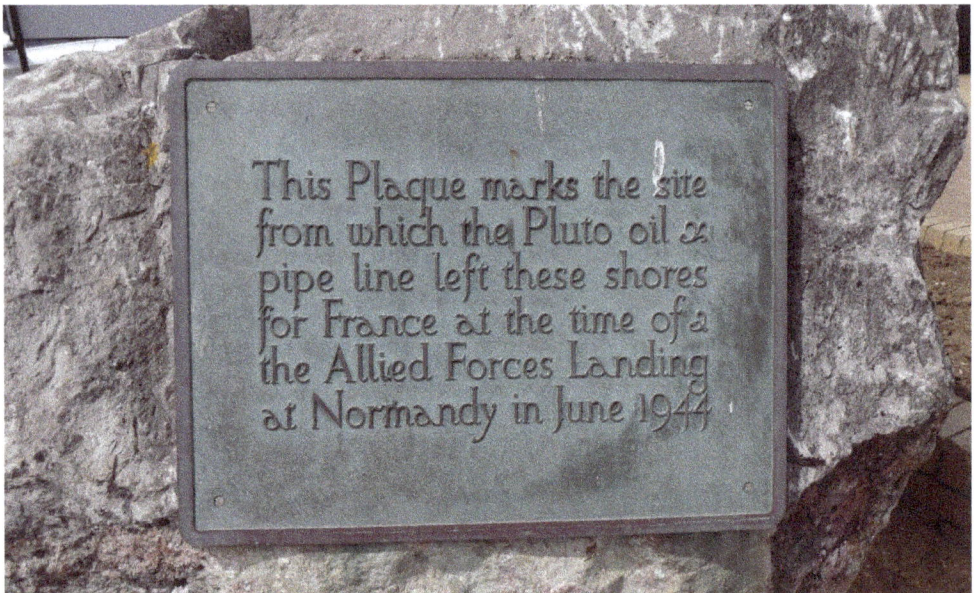

The first two successfully laid Cross-Channel **PLUTO pipelines once ran out along the 1,200-foot pier behind you.** They drooped across the gap cut as an anti-invasion measure, and they then plunged off the end to travel 72 miles across the sea bed to Cherbourg. The pier was closed in 1975 but reopened in 1976 under the ownership of Fred Sage. Shows and cruises resumed and the Shanklin Pier Preservation Society was formed to raise money for repairs. Leading Leisure plc took over in 1986 but on 16th October 1987 a huge storm destroyed sections of the deck. Eventually South Wight Borough Council bought the pier for £25,000 and then spent £189,000 demolishing it in February 1993.

*The Royal Spa Hotel, Shanklin. Above 1928, below taken before the lift was built in 1891.*

*Shanklin Pier c. 1945. The gap cut to prevent its use during any invasion is clear. If you look carefully, there is a possibility that it caught the PLUTO pipes draped across the gap.*

Recent research indicates that under your feet was a reinforced room, used as a kiosk through the 1950s, which was most likely used as the manifold room to connect the pump bus ring main pipeline to the sea pipes bringing the pipes under the road, invisible to prying eyes from above. If you look to your right, with your back to the sea, to the right of the café there is a much heavier groyne along which ran the two pipelines that stretched across the bay to the Sandown pumping site, with the Granite Fort clearly visible 2.6 miles away.

Work on the Shanklin pumping sites started in the summer of 1943 with the construction of the pump house buildings. The Shanklin site consisted of eight reciprocating pumps numbered 19-24 and 26-27 and one centrifugal pump numbered 25. The erection of the pump houses was due for completion by January 1944. However due to the slow delivery of materials and lack of adequate workforce timescales had slipped but by March 1944, construction of almost all buildings at Sandown was complete and by the middle of May, most of the work at Shanklin had been completed. By the beginning of July, the pumps at Shanklin had been water tested and by 5th August the pumps had been spirit tested and certified ready for operation. Caterpillar diesel engines were installed to provide power for the reciprocating pumps, while the centrifugal pumps were powered by electric motors, connected to a high voltage electric supply, specially routed in by the Isle of Wight Electric Light and Power Company to an adjacent transformer house. Valve pits were constructed between the pump house buildings in the vicinity of the cliff lift and camouflaged by covering them with suitable rubble effect mesh. The Shanklin pumps were housed in purpose-built buildings, hidden amongst the bomb damaged houses and hotels along the seafront in front of you. Before the war, the buildings along the seafront between the lift and the present day amusement arcade were all apartments, rented out to holiday-makers.

If you cross the road toward the cliff lift a pump house containing pumps number 23 & 24 was situated near the foot of the lift. This building was thought to have been demolished to give access to the new lift, but a current theory suggests that it may have been in the toliet building againt the cliff - possibly also the location for the PLUTO control room. All pumping plant, such as valve and strainer pits and all pipework was removed post-war.

Behind you, to the rear of the Sunny Beach apartments pumps number 21 and 22 were located in a purpose built pump room. This has now been incorperated into the apartments but you can make out where it was located. Nearby, about 100 yards to your right two reciprocating pumps, numbers 19 & 20 were located in a single storey white building behind what is now the Shanklin Beach Hotel. Before the war, this was a semi-detached building, one side of which was called Ocean View. This PLUTO pump building still exists as additional accommodation for the hotel.

*PLUTO Pump House, Shanklin.*

*Fully camouflaged.*

Photograph of the same (lower floor) PLUTO pump house behind the Shanklin Beach Hotel, taken from the top of the cliffs.

Patch and Judith have found the PLUTO pump house!

*If you know where to look clues to PLUTO are all around. Here the scars of two original vents can be seen.*

It is possible to walk behind the hotel through the car park (accessible from the main road) where the original single storey building housing the pumps now has a second storey added containing hotel rooms. Blocked doorways and openings are clearly visible and this is the only surviving PLUTO pump house in Shanklin.

The dimensions of the new pump houses seem to have been the same. This surviving building has a length of 55 feet, a width of 23 feet and it stands about 10 feet high. There are vents, measuring 4 feet by 1 foot, at the foot and head of the walls at various intervals. These vents were to provide sufficient ventilation in order to eliminate any danger from petrol fumes. The walls were constructed from brick with a flat cast 14-inch concrete roof. The roof was made out of rectangular concrete units (about 5 feet by 18 inches) with a concave under-face. These were laid side by side on iron joists laid horizontally across the width of the building. Inside, a concrete blast proof partition wall divided the engine room from the pump room. Inside renovations have hidden all trace of the ground floor's pumping history although the author was able to visit during works there and the original cast concrete arched roof structure still survives.

On the pump house roofs, fake bomb-damaged walls and imitation doorways and windows were built to simulate bomb damaged rooms. Part of a fallen staircase was even erected on the pump house behind the present day Shanklin Beach Hotel. Attention to detail was paramount, so even cavity walls were constructed and an irregular sloping wire mesh platform was built off the side of the roof and strewn with rubble to simulate a pile of debris leaning against the wall of the building. Above the pump house next to the cliff lift, a shattered, hipped roof was constructed with small gaps between the tiles to simulate a roof shaken by a bomb blast. The pumping site would have been invisible from the air.

Walking back past the pier and the old Royal Spa Hotel car park a pump house containing pump 25 has been lost somewhere under the 'Napier' block of flats, possibly where the garden now is. Today nothing is visible.

Slightly further on, on the seafront, two more pump houses containing pumps number 26 & 27 sat in buildings that may have originally been garages where the *Coles Bazaar & Snackery* now sits. After the war the pumping equipment was cleared out and the buildings went back to being tourist facilities.

Very few photos exist of PLUTO on the Isle of Wight as all the beaches and sea fronts at Sandown and Shanklin were out of bounds and heavily fortified. In his book 'PLUTO: PipeLine Under the Ocean', Adrian Searle includes the memories of Reg Langstaff, from No. 13 Company, Pioneer Corps, whom he had interviewed about his work on the pump houses at Shanklin. He recollected:

> *' - at Shanklin, a huge concrete-mixer was installed inside the ruins of the Royal Spa Hotel. Each time a lorry entered the building to load up with concrete etc., the doors immediately closed behind it. The same thing happened after it left to take the load along the seafront, and then behind the hotels, where buildings were erected to house the pumps. Some men were detailed to brush out the tyre marks which were made each time a lorry went along the front - in case enemy aircraft came over to take photographs. Everything had to be maintained as if nothing was taking place.*
>
> *I remember the workings of Shanklin Cliff Lift were removed to make way for the project, and I guess there may have been a shortage of timber during that period because we were detailed to strip quite a few of the hotels along the front of all their timbered floors. The wood was then used to shore-up the solid concrete 'bed' inside the buildings which had been put up behind the hotels. Once all the floors in the hotels had been removed, you could see the sky through those parts of the roof where slates were missing, as you looked up from the bare ground below'*

Captain Havinden was in charge of camouflaging and had to liaise with Headquarters Tactical Air Force to make sure regular aerial photographs were taken on a *'fortnightly cover'* basis and a 'camouflage report' accompanied the progress reports. These could be studied to make sure there was no trace or visible irregularity that might give German Intelligence a reason for further investigation. In Shanklin pipelines were routed as much as possible along kerb lines, so that it blended in with the kerb. At other places, *'Camouflage netting'* was provided to hide pipes and pits.

*This (now toilet) building sits to the rear of the car park where the Royal Spa Hotel once sat. In is thought to be WW2 and its construction techniques match the other surviving pump rooms. Was this the 'lift pump room' and being next to Osborne steps possibly also the 'Osborne PLUTO Control' room.*

*If you walk further past Coles Bazaar & Snackery* **(where PLUTO pumps once stood)** you will arrive at the bottom of Shanklin Chine, but there is no trace of PLUTO to be found. As a side note it is known that the military took over a large building on Keats Green above the beach area in Shanklin where the PLUTO pipes were welded together by joining them with high voltage welding machines. At the moment we can't identify which building it was but it is likely to be the Clifton Guest house or one of the large houses on either side.

# 22. SMALLBROOK AND RYDE SANDS

## *TWO SHORT VISITS TO FIND TOMBOLA*

*PLUTO TOMBOLA pipes at the entrance to Ryde Harbour.*

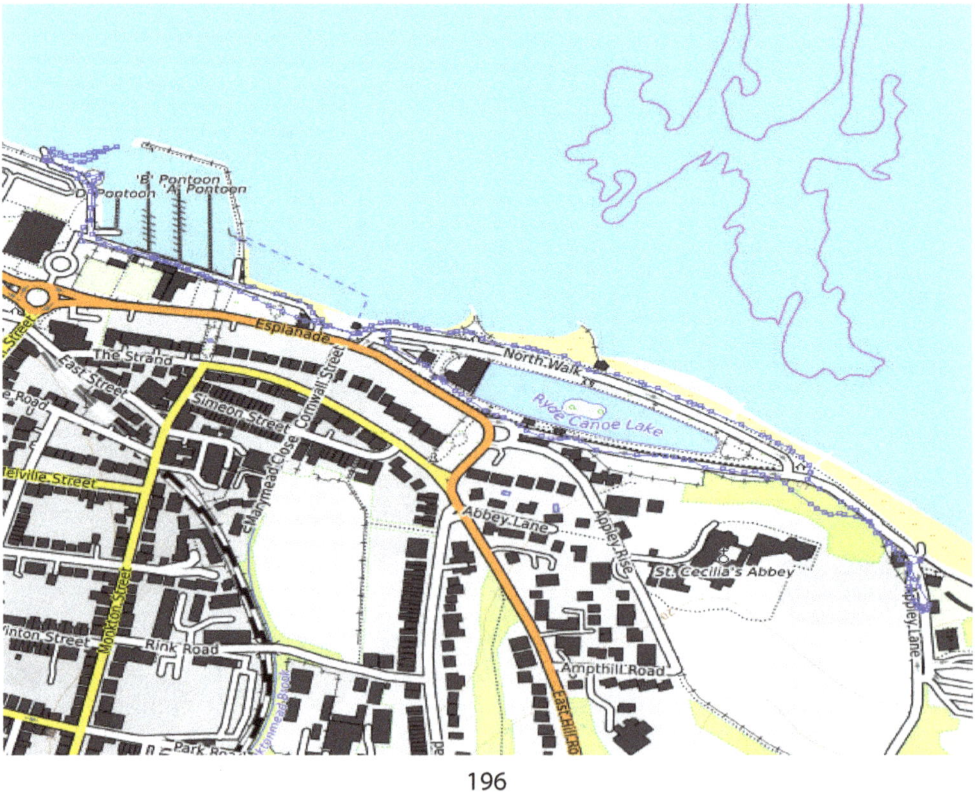

Some of the earliest trials for TOMBOLA were undertaken on the sands of Ryde beach.

One option is to just park behind the Ryde SuperBowl bowling alley, however to make more of a walk of it we suggest parking further up past the Canoe Lake, there's plenty of parking – the map starts behind the Three Buoys Cafe; follow the Esplanade to the harbour, past the pedestrian entrance to the harbour wall, and continue to walk toward the harbour mouth entrance (for boats!) and look across (at low tide) to the foreshore, **you will see three PLUTO pipes.** These were dredged up out of the harbour entrance a few years ago where they were becoming a hazard and were then just dropped onto the bank. If you walk right around the harbour, outside the breakwater, you can scramble down to them, but it is slippery. Directly crossing the harbour mouth is slippery, and you WILL sink in the mud. As you walk, the beach to your right is where TOMBOLA and PLUTO minor were designed and tested and the men trained to assemble the system in France after D-Day.

John Sullivan remembered:-

'East of the 2200-foot train trestle to the ferry dock at Ryde was Ryde Sands. When the water went out for low tide, the beach was quite wide. This is where the training by the British Army pioneers was done for their pipeline ship-to-shore practice.

We all went to observe the building of a ship-to-shore line, or as we then called it, the sea line. The procedure was to place a board foundation with rollers every ten feet, and alongside each roller, place a board foundation, and erect an edge of a 2x6, perhaps 10 to 15 feet long aside each roller, attached to each erect 2x6 was a flat board underneath to spread the weight as a footing. On each one of the upside down T, we would lay and connect pipe, 500 feet strips apiece, alongside the rollers, and then connect all the 500 feet lengths desired to meet the depth needed for a tanker unloading at ebb tide.

The pipe was approximately 6 inches inside diameter and 6 ⅝ outside diameter. A flexible hose was connected to the pipe at sea with a float at sea so the unloading tanker could find it. The pipe, when pulled out to sea by a trawler, was attached to a 'sea-sled' and then the next 500-foot pipe was connected. They were quite proud of the fact that they could average 1329 feet in an eight-hour day of connections. The only problem was that the tide was only out for a little less than four hours. Each 20-foot piece of pipe weighed 455 pounds (206Kg). Eight men were carrying the pipe from the dry beach pipe stack, to the work site in the low tide area. They had at least 150 to 175 men working at the work site. We would have a British Navy trawler pull the 500 foot lengths to sea, and connect the next 500 foot lengths until the desired length was reached. At least a half mile of sea line was contemplated to get to the depth a tanker would need to float.'

Roger Bartrum remembered: 'The pipeline at Ryde was around 200 yards east of the pier and disappeared into the Smallbrook stream outfall'. He assumed it linked up with PLUTO in Sandown Bay, but it most probably ran to the Smallbrook tanks.

## THE SMALLBROOK TANKS

At a PLUTO lecture a few years ago, one of the audience spoke to me about the PLUTO tanks at Smallbrook. About which I knew nothing!

During WW2 his grandfather owned the wood to the south of Smallbrook Lane and east of the Smallbrook train junction. There is a small area where a car can be parked just by the humped railway bridge, where the footpath starts. This wood was compulsorily taken over in 1943 by the military and the entire centre of the wood was cut down.

Large tanks were then constructed, the PLUTO team working day and night. The owner and his family were told that it was top secret work and they must stay away. But his mother, as a child, remembered the work well, including a visit from a very high ranking officer who she thought might have been Admiral of the Fleet, Louis Mountbatten.

Careful study of the c. 1945 photographic images does show that the wood has been 'hollowed out' and also shows at least three distinct rings (one of them very large) and other disturbed areas. These are thought to be TOMBOLOA storage tanks. It is not clear if these were just being fabricated on site ready for D-Day or whether they were installed there and a pipeline was run down the Monktonmead Brook (or alongside the train track) to support operations and even fuelling for D-Day from Ryde Beach. Mr Bartrum stated that a pipeline did come out of the Smallbrook stream onto the beach, while another source refers to 'the PLUTO tanks' located in Whitefield Woods, perhaps half a mile from the Smallbrook site.

Today there are only pieces of random brickwork and concrete left in the ground, but you are **undoubtedly walking where PLUTO was based.**

Of course the Ryde pipes and the Smallbrook tanks are something of an anomaly. They had no known feed from the main PLUTO pipes, but rumours persist of links across the Island and in the conclusion section we offer a theory that the main PLUTO pipeline was extended from Sandown, along the railway track to feed into a much larger (i.e. Hungerberry size) tank and then onto Ryde beach.

Another suggests that the pipelines constructed on Ryde beach were solely for training, so that the engineers were ready for rapid construction in France after D-Day. Hence the Smallbrook wood facility may just have been for manufacture of tanks ready to be shipped off to France. Although several references can be found to a pipeline that 'dissappeared' into the smallbrook stream off Ryde beach.

From D-Day +15 the plan called for tankers to berth at Port-en-Bessin and discharge bulk supplies into tanks constructed on the quay side. Small tankers could berth inside the outer harbour and larger ships would anchor 1,000 yards outside the port and submerged pipelines connected to both ship and shore. This system was called TOMBOLA and these were available on D-Day +9, storage tanks being sited near the Vauban tower, the pipes coming ashore on the western harbour wall.

Details of the tanks are difficult to find. The tanks for the main Air Force Reserve Depots (AFRD) were about 4,000 tonnes in capacity, known as the C2 design. This was a vertical, cylindrical, reinforced concrete, semi-buried, steel-lined chamber built on a concrete foundation. The concrete roof slab was supported on 45 columns within the tank. It was relatively cheap to build and proved excellent in service.

These PLUTO tanks were much closer to the C1 type tank, built in the same way as the C2, except that it was above ground. They were only usually constructed in areas where the risk of air attack was considered to be very low. Tanks built up to the 1930s had been riveted, but one the most remarkable thing about the whole PLUTO tank programme was the reliance upon welding for the pipes and the tanks.

The programme for building protected storage fuel depots across Britain had proceeded at an increasing rate. Greater mechanisation and more experience meant that the rate of tank construction improved. It took 4½ weeks rather than 6½ weeks to build a C2 tank, and the required labour force fell from 30 to 20. Most large commercial installations were camouflaged, and bunds were built, where possible, around the tanks.

# 23. FROM THORNESS TO REW STREET –
## *A LITTLE STROLL ALONG THE BEACH* (About 2.3 miles)

Whippance Farm is on the far right of this picture. PLUTO ran down the hedgerow and around the curved hedge - or rather the hedge followed the pipe from the pumping station just to the left of this picture and the path.

For those who feel like a walk down the beach (best taken around low tide) there is another intriguing part of the PLUTO system hiding round the corner from the main PLUTO manifold and the must-visit PLUTO pipes that stretch out across the sand and mud at Thorness.

Probably the best place is to find your way and park at the end of the road that leads down to Thorness beach accessible through Thorness Bay Holiday Park – this is a bit of a one-way system, and the beach lane gate may shut at night. But follow your nose (clue - follow the signs for 'Thorness Bay Area') down to the beach where there is plenty of parking.

From there, head north around the beach toward Whippance Farm and you will eventually arrive at the main PLUTO pipes. **Here PLUTO definitely crossed the beach!**

The remains of the manifold frame still survive along with 12 or 13 pipes (of the original 20) that at very low tide still head out to sea toward Lepe on the mainland for perhaps 50 or more yards. Be careful as the Blue Slipper mud is both slippery and very sticky at the same time.

If you look behind you up to the bank top, that is where the PLUTO pipeline ran, back towards the roadway to the farm. At the base of the modern plastic water chute you may be able to find **a group of dumped PLUTO markers in the weeds**, we assume pulled out of the nearby hedge rows.

*Discarded, but not forgotten.*

The pumping station that pushed fuel across the 14-mile Island single pipeline once sat in the (often flooded) area just off the beach, and before the roadway from the beach to the farm. The PLUTO pipe then arched to your right, **crossed the path,** and followed the curved hedge line until it then again **crosses the path near the farm.** The walk now continues to walk around the beach, toward Rew Street point.

But if you would like to divert up the hill taking the coastal path, on the headland you will find the remains of one of the Island's many large Heavy Anti-Aircraft batteries (HAA). Here four 3.7 inch AA guns once each threw into the air 17 rounds a minute of high explosive shells to protect Southampton, across the water. Pieces of shrapnel and shell fuzes can still be found across the beach as you walk. Today two of the Rew Street HAA four gun positions and their surrounding ammunition storage bunkers survive and further inland the gun's radar control room and barracks, complete with a blast wall are all intact and can be accessed. Some of the (private) holiday huts are rumoured to have been left over from the battery.

Should you take this excellent detour then you will have to return back to the pipe manifold to carry on round the beach, as there is no easy access down from the coastal path to the beach. As you walk along the beach you will start and see large sections of PLUTO pipeline – not the 3-inch steel HAMEL pipe, but 8-inch main PLUTO pipeline. Eventually you will come to the foundations of a small square building. From this point, **PLUTO crossed the beach**. Aerial photographs show two pipelines running out to sea. We suspect that these were TOMBOLA refuelling pipelines – fed directly from the main Thorness manifold.

At the end of the sea pipes, flexible pipes would have refuelled the tankers heading to the Normandy beaches. You can walk on to Gurnard, but the route back will get you to your car – but please watch the tides!

*A new part of the PLUTO story has unfolded on the beach to Rew Street. Long stretches of 8 inch diameter main PLUTO pipes run round to the foundations of a brick building where pipes still run out to sea. Another TOMBOLA site has been identified, half way between the main manifold and Rew Street. Look just beyond the remains of the blocks from the collapsed HAA battery from the cliffs above. A brick frame was another valve room and again aerial photographs show another pipe running out to sea.*

# 24. A STROLL AROUND BROWN'S GOLF COURSE

## AND THE PLUTO PAVILION IN SANDOWN

The best preserved parts of the PLUTO project on the Isle of Wight are scattered around Brown's Golf Course in Sandown Bay. This stroll starts in the car park at Brown's Family Golf and Café, Yaverland Road, Sandown, Isle of Wight, PO36 8QA which offers two golf putting courses (18 and 9 hole)- and an excellent tea room!

Brown's golf course was designed and opened in 1932 by professional golfers Henry Cotton and Joe Kirkwood. Cotton achieved fame during the 1930s and 1940s with three victories in The Open Championship (1934, 1937 and 1948) and he placed 17 times in the top ten at the Open. His record round of 65, made during the 1934 Open Championship led to the Dunlop golf company issuing the famous 'Dunlop 65' ball. Cotton also succeeded in winning many titles on the European circuit during the 1930s. Joseph Henry Kirkwood Snr. was a professional golfer who is acknowledged as having put Australian golf on the world map. Kirkwood's best performance in a major championship was a third-place finish in the PGA Championship in 1930, a semi-finalist in the match play competition. He finished fourth in the British Open on three separate occasions and in 1933, he won the Canadian Open. In 1936 this main golf club building was rebuilt (for the third time) and this is the building that still stands today - as Brown's cafe.

This walk starts in the glass roofed Brown's tea room.

**You are now standing where a PLUTO pump once stood,** installed in 1944.

For a long time, the vulnerability of the original glass roof of this building during war time concerned me, but the term 'hidden in plain sight' was finally explained when an eye witness explained that her father had been one of the carpenters who had built a complete false wooden roof inside the glass roof. The top side of this was painted green and textured to match grass. The camouflage artists even added three dimensional picnic tables and chairs. So soak up the atmosphere, have some tea and then walk out of the building onto the golf course. To your right is the Brown's Ice cream factory building, built in January 1938. The factory closed on the outbreak of war in September 1939, and the golf course was soon converted to vegetable plots as part of the 'Dig for Victory' campaign in 1940.

From here you can go in two directions. You can turn left and visit the Power Pavilion and then walk around the golf course or you can turn right and walk over to the (Ice Cream Factory) building on your right. In this building the room nearest the sea **contained a PLUTO pump,** and still has its heavy lifting rail mounted in the roof. Each pump weighed around 7 tons and had to be moved and installed by hand. (Please note - both buildings are today private offices.)

This PLUTO control room has often been reported as having been located in the basement of the Grand Hotel, the now not so grand building between the Ice Cream Factory and the Granite Fort. However recent research would point to this possibly having been located at the front of Brown's café, where you now hire golf clubs. Another option was that it was in the rear of the

Ice Cream Factory building in front of you, although I now suspect this was the communications room for the PLUTO pumping site.

Walking past the Grand Hotel, where many troops and later PLUTO engineers were billeted throughout the war, you will come against the great walls of the Sandown Fort.

This is one of the many Palmerston Forts built on the Island in response to a perceived threat of French invasion. In 1859, the Royal Commission felt that earlier defences did not offer suitable protection, so construction of the fort began in April 1861 and was completed by September 1864 at a cost of £73,876. In later documents it is often referred to as the Granite Fort. The fort originally had eighteen 9-inch Rifled muzzle loading guns facing the sea behind iron shields, these guns were later upgraded and an extra five inches of armour was added. The French invasion never came, and by the turn of the century most of the Palmerston Forts on the Isle of Wight were disused, or simply being used for storage.

In 1930 the fort was sold off and demolition for building material started, removing the entire rear part of the fort. But with the advent of war in 1939 the fort was recommissioned to protect against a German invasion that was always threatened to arrive in Sandown Bay.

Today the fort houses the Isle of Wight Zoo, created in the 1950s. During that decade Brown's probably reached its zenith in terms of popularity when the attraction extended across almost 20 hectares and included its own ice-cream factory and the adjacent Canoe Lake and bakery, which produced much sought-after cream doughnuts. But today, back towards the main town the Dinosaur museum is a new addition to the local attractions.

Today Sandown Zoo is a great place to visit and apparently has many exciting animals inside. But of course for the purposes of this walk it's the PLUTO story that has brought us here and the amazing fact that in 1944, 14 (of the 16) enormous PLUTO pumps were housed in what was then a derelict fort.

Today the zoo also has a World War II themed display that tells the story of its unique part in wartime history. It also has the only PLUTO pump in an original position. (Entry fee applies.)

Up until 1998, it was believed that only two pumps had survived - one in the Bembridge Heritage Centre and one in the Imperial War Museum. However, another was found being used by the National Grid to wash electrical insulators and this seven-tonne pump was brought back to its original position under the fort's gun arches. The official opening of the display took place on the 66th anniversary of D-Day, on Sunday 6th June 2010.

Of course there are some mysteries at the Sandown pumping site. There should be pipes leaving the fort and it would have been assumed that a large ring main pipe would have linked all the pumps along with many manifolds, filters and valve. The pipeline that looped around the town of Lake from the main Hungerberry Wood tank must have come into the back of the fort and split to the café and ice cream factory pumps. Somehow the two pipes that came across the bay from Shanklin must also have joined into this system.

But there are no known plans or even photographs of what should have been a major installation and today there are no signs of any of this pipeline and no visible scars or sealed openings that can be found. It is known that the Sandown site was for many months considered to be the 'A' site for PLUTO, but in the end the two Cross-Channel pipes went into Shanklin.

***The question is, was Sandown ever made operational?***

One key part of the PLUTO project was to ensure fuel supply through duplication. The project had to ensure that there was sufficient redundancy in the system in case any part of it was destroyed by the enemy. The Sandown site was undoubtedly the backup site for Shanklin. We now believe that it was never completed, probably due to time and material constraints. But should the Shanklin site have been damaged by enemy action, then Sandown was available, and could have been brought online perhaps within weeks.

There is one more part of the PLUTO Project here at Brown's golf course to visit. Across the golf course there is a wooden Pavilion, that over recent years has become the source of some controversy. To reach the building it is easiest to backtrack past the café and follow around the edge of the golf course to the edge of the larger car park, avoiding the golfers and the greens. In the corner of the wood volunteers have built a very pleasant woodland walk to take you to the building.

*A common misconception exists that during 1944 this Brown's Golf Course Pavilion housed a PLUTO fuel pump. This was never the case.*

Now known as the Brown's Golf Course PLUTO Power Pavilion, the original hut was built in 1936 to supply DC electrical power to the golf club behind you.By 2014 the Pavilion building had fallen into a significant state of disrepair and the decision to restore it (externally) was taken, in part aided by a grant from the Coastal Revival Fund. The structure has been a Listed Building since 2006, designated as a DC power generation building for Brown's Golf Club.

A building is listed when it is of special architectural or historical interest and considered to be of national importance and therefore worth protecting. As the term implies, a listed building is actually added to a list; the National Heritage List for England. Listing is not a preservation order, preventing change and it does not freeze a building in time. It simply means that listed building consent must be applied for in order to make any changes to that building which might affect its special interest. For failure to comply the maximum penalty is two years' imprisonment or an unlimited fine!

Inside the Pavilion lies an historic treasure trove of period electric generators, motors and a complete 1930s electrical control and distribution board. This alone would have justified its Listed Building Status, but its long, if somewhat vague association with the PLUTO project obviously added to the building's interest. The original 2006 listing suggested that the pavilion provided power for all the PLUTO pumps (which was quite clearly impossible). (It also referred to the control room in the Grand Hotel which is also now not thought to be correct.)

Naturally the essence of any historic research should be to update the record as new information comes to light and listings can, of course be corrected.

But in 2016 a concerted effort was mounted to get the building 'delisted'. The contention was made that the Pavilion was simply a building housing water pump(s) to water the adjacent golf course and may have even had no part at all during WW2. The Pavilion's only purpose was to power two pumps to water the golf course greens, the clubhouse lighting and the adjacent ice cream making machinery. Considerable emphasis was placed on the fact that there is no hard evidence to support a PLUTO link in the nationally archived PLUTO documentation. Conversely there is nothing that disproves it.

Remember right at the beginning we said that: 'Absence of evidence is not evidence of absence'; Such that the lack of evidence to prove something, does not disapprove it. Indeed, the paperwork and evidence about Pluto on the Isle Wight is, at best sparse.

But delisting a listed building is also an exceptionally unusual process, only done if the listed building has already been destroyed or is in the way of major development. In this case delisting the Pavilion would remove all protection from future development of the area or indeed destruction of the building. Rumours started to fly as the area had become the focus of a proposed major housing development project and of course delisting the little Pavilion could open that door.

Major submissions were made on both side of the argument, an Historic England inspector visited and the committee made their decision. The Pavilion was indeed part of the PLUTO project, albeit in a far different capacity. The listing status was confirmed and rewritten.

Predictably even this didn't quieten the doubters who then filed an appeal! This was also unsuccessful and Historic England unequivocally dismissed the application (and appeal) to de-list the Sandown pavilion and accepted the validity of the revised case. The future of this amazing building and it history has been protected for future generations.

Despite (and during) all this nonsense in 2016, a group of mostly volunteers, started work to save the building.

The power pavilion was built in 1936 to produce DC power for the golf club, now the café where this walk started. Indeed, the construction techniques of the two buildings are identical including the herringbone brick porches and the elm boarding to both fronts. At this time (1936) the Bakelite distribution board, fuses, dials and knife edge switches were put in place.

Since the main building restoration project was completed in 2017 and the structure was made secure and watertight, a dedicated team from 'Men in Sheds' have taken on a painstaking restoration of the engines and equipment inside. They believe that the engines can and will run again – a superb effort. They have regular open days, accept school trips and are pleased to show off the increasingly well preserved and soon to be working engines and equipment - so please support them!

**Do we know what the PLUTO Pavilion's exact purpose was?**

No, although significant research is ongoing and now the building is safe we will continue trying to understand what actually happened there!

From the evidence amassed it is clear that the 1936 Brown's Golf Club Power Pavilion was significantly rebuilt during WW2. Significant steps were taken to disguise large underground fuel tanks and a system was installed to bury, muffle and mask the engine exhausts.

During the short life of the PLUTO project in 1944 its use as a power station enabled it to provide the power supply for the Brown's cafe building (which was powered by DC pre-war) and which of course housed a PLUTO pump during 1944. All the cabling would have been in place and no modifications would have been required to the building. For a project under huge time pressure this would have provided a very useful saving in time, material and labour.

It is also highly likely that the Power Pavilion provided backup for the main Sandown Pumping station control centre, the communications room (likely to be DC), the Granite Fort and its defences. Indeed, if it can be proved that the PLUTO Sandown communication room was located in the front of Brown's golf club, and the control room to the front of Brown's Ice cream factory then the Pavilion would have been required to run full time during PLUTO operations. There are other examples of similar power generation use during WW2 using DC motors.

During the war the Pavilion's main strength was to hide in plain sight. By 1942, building camouflage had become restricted or simplified to painting in greens, browns or blacks. In some cases, texturing of roofs was effected by applications of either stone or wood chips on a binding medium. As restoration progressed in 2016, wartime bituminous emulsion on the roof was found.  The team also found that the front fascia of the Pavilion was painted in a military green. Samples taken from door hinges indicate that this was BS381C: 1944 - Colours for Ready Mixed Paints - NO 25 LT Brunswick Green. During war time the golf course was out of use and there would have been no reason to paint the building or repair the roof. Indeed, there would have been no access to the building for the owners during most of WW2. Today it has been restored to its wartime colours.

Captain Lickens, who was in charge of the BAMBI PLUTO installation during the Second World War recounted an inspection carried out at Sandown before D-Day, in which he was made aware of the alterations made to the pavilion in support of PLUTO. Maurice Lickens would receive a mention in despatches for his PLUTO work, and was awarded an OBE for services to Isle of Wight tourism in 1997.

Another interesting part of the PLUTO story came to light when a series of documents were found stating that in 1944 the pavilion was guarded by the elite and highly trained Home Guard Auxiliaries Patrol, both before and after the D-Day invasion. You don't randomly guard a building on the edge of a golf course during D-Day for no reason.  In addition, the recently discovered PLUTO 1944 Fire Alarm response documents, specifically list the Power Pavilion.

At that time and that place the only reason to protect the building was PLUTO. **As you stand in front of the Pavilion you are standing in front of PLUTO. To** return to the café you can walk around the golf course, we suspect that the PLUTO feed that came around Lake, looped aorund the edge of the course to enter the (demolished) rear of the fort. Or you can walk back the way you came and treat yourself to a Cream Tea!

*Inside the PLUTO Power Pavilion.*

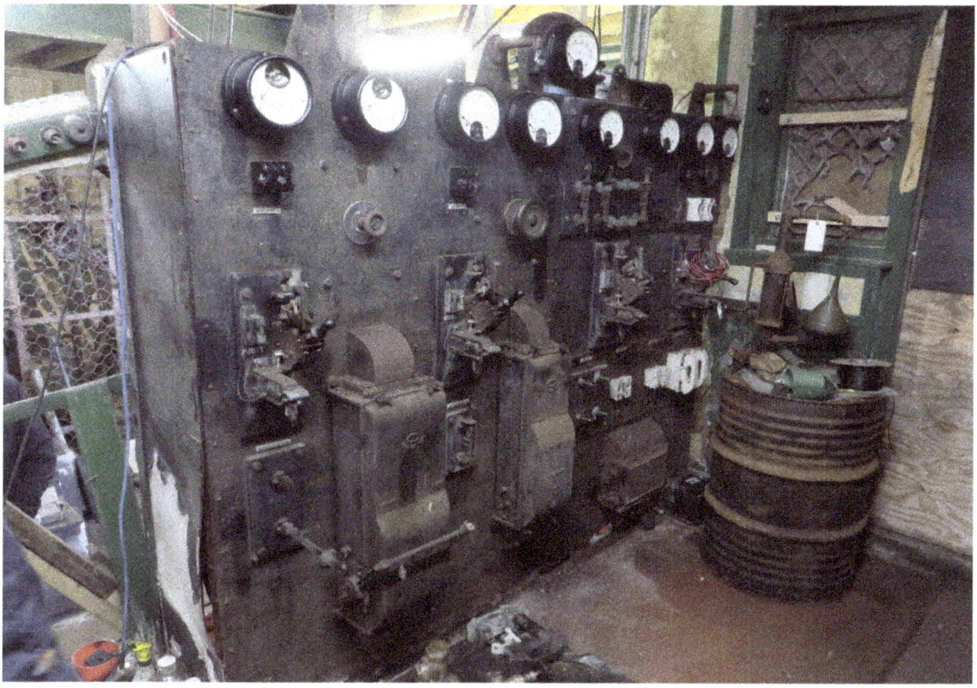

# Where PLUTO Crossed the Path

*Rambles with a Purpose*

*on the Isle of Wight*

SECTION 2 - THE PLUTO STORY

CHAPTER 2.1

**D-DAY 6TH JUNE 1944**

*There have been many days when the actions of brave men change the world.*

Seventy-five years ago, on June 6th, 1944, the largest seaborne invasion in history took place on the beaches of Normandy. Allied troops began to invade France in an operation that would eventually culminate in the end of the Second World War.

On 31st May 1944 troops were moved from their temporary encampments to begin boarding landing crafts and ships. All communication and 'fraternization' with the locals was forbidden. Along the South Coast there were twenty-four designated embarkation points, all with troop marshalling areas close by. There were four embarkation points in Portsmouth and three in Gosport. The Eastern Docks of Southampton were used for docking the larger ships and the Western Docks sheltered landing craft. Southampton Town Quay had three separate embarkation points for troops boarding landing craft. Both the Hythe Ferry Terminals and the Ocean Cruise Terminals were full.

On the Isle of Wight, Cowes, Newport and Ryde were packed and the Island refuelling stations at Rew Street, Cowes and Ryde and the degaussing station at Puckpool were fully committed with long queues of vessel forming. The Eastern and Western Invasion Task Forces started to assemble with literally thousands of Allied combat ships of every description filling Portsmouth Harbour and spilling out along the Solent from Southampton Water to Chichester. All were silent, waiting for the dawn.

The invasion fleet, codenamed Operation NEPTUNE, consisted of five great forces, one destined for each landing beach on the French coast. In five principal convoys, the force contained more than 5,300 ships of all types, 4,126 landing craft and 4,000 relay boats between the shore and the ships. It was the largest single amphibious invasion of all time. Admiral Kirk led the American sector: Force U (for UTAH beach) based at Plymouth, and Forces O (for OMAHA beach) based at Portland. The British, Canadian and Free French sector was led by Admiral Vian's Force S (for SWORD beach) based at Portsmouth, Force G (for GOLD beach) based at Southampton, and Force J (for Juno beach) was based on the Isle of Wight. Additional support forces (Forces B and L) were based close to Falmouth and the Nore while 12 minesweepers constantly cleared the convoy channels.

Each of the naval convoys started moving at different times according to their location all planned to meet at a place code named 'Z', but more widely known as 'Piccadilly Circus', precisely 30 km south-east of the Isle of Wight.

From there they moved towards their respective beaches, supported by a massive naval bombardment pouring high explosive shells over their heads to defend the landing craft as they met the beaches and the defending forces. The Allied support armada comprised 325 warships, that included 101 destroyers, 6 battleships, 2 monitors, 22 cruisers and 93 destroyers.

D-Day was originally set for the morning of the 5th June, but a large storm hit the English Channel on Saturday 3rd June and the whole plan had to be postponed for 24 hours because of the bad weather.

On the night of 5th June Islanders remember being woken by the sound of thousands of aircraft flying overhead. The invasion of Normandy had begun. From 11 p.m. on 5th June some 24,000 airborne troops were dropped behind German lines to secure important roads and bridges. Along with more than 2,000 aircraft some 867 gliders were deployed. Sixty-eight

Horsa gliders, carrying men of the 6th Airborne Division, along with four giant Hamilcars loaded with heavy equipment, headed for their landing zone. Forty-seven Horsas and two Hamilcars reached their destination. As daylight took hold, another two hundred and fifty gliders, most of which were Horsas, delivered 7,500 men directly into the battle zone behind the beaches.

D-Day, 6th June 1944 saw some 160,000 soldiers, Americans, British and Canadians land on five beaches spread across 50 miles of the Normandy coast. It took almost 200,000 Allied naval and merchant navy personnel to get them there in some 5,000 ships and support vessels. In the next few weeks, Allied forces would land more than 2 million men in Normandy.

When civilians awoke on the morning of 6th June all the military vehicles and personnel had vanished overnight. The morning was like no other before or since, and those who remember it will never forget that day. Everyone was up early, but most had not slept at all. The sky had been full of aircraft and the Solent had been a wonderful sight with line upon line of ships stretched from Yarmouth to Hamstead and then a further similar formation from Salt Mead buoy to Cowes. Witnesses thought that you could walk from the Isle of Wight to Portsmouth on the giant armada without getting wet. Now they had all gone.

At the Bembridge Fort, Chain Home Low radar station and Culver Wireless station the night duty staff knew that 'this was it'. There was no time to return to their 'digs' in Sandown. They cooked their own food from a supply of thinly sliced meat with gravy and potatoes over a gas burner. The fried bread came up well and was very crisp, and was subsidised with a few eggs from friendly farmers while they all waited for the dawn.

The D-Day landings were made in two stages. Shortly after midnight British, American and Canadian troops were landed by glider and parachute to seize key objectives such as bridges and road crossings. At 6.30 a.m. infantry and armoured divisions were landed by amphibious vehicles along the Normandy coast. The logistics are staggering. Nearly 7,000 assorted ships, including over 1,200 warships began the invasion with a dawn bombardment. By the end of that day 160,000 Allied troops had been landed, and the offensive against the occupying German forces had begun. By the end of June this total had swelled to over 875,000 men, fighting to liberate Europe.

**D-Day would become the Allied forces' longest day.**

Estimates put the Allied cost of D-Day at 4,414 confirmed dead, nearly half of those at the heavily defended Omaha beach.

But the Allies took all five beaches on the first day, and later linked them into a perimeter that eventually paved the way to the Allied victory on the Western front. With the Russians advancing from the East, it secured the liberation of Europe from the scourge of Nazism.

# Where PLUTO Crossed the Path

*Rambles with a Purpose*

*on the Isle of Wight*

SECTION 2 - THE PLUTO STORY

CHAPTER 2.2

*THE NEED FOR FUEL*

*My boys can eat their belts, but my tanks gotta have gas!*

General George S Patton.
Normandy 1944

D-Day required an enormous planning effort. As the invasion of mainland Europe progressed Allied commanders knew that as the German army retreated they would destroy every fuel storage compound as they left. They also knew that as the Allied troops advanced through Europe during the first few weeks it would become increasingly important to ensure a reliable source of fuel for all their vehicles.

So an ingenious, audacious, incredible, almost ridiculous plan was put into action to meet this need.

It was called **PLUTO: Pipe Line Under The Ocean.**

As the enormous invasion force sailed past the Isle of Wight in the early hours of June 6th 1944, the incredible PLUTO project, at least on the mainland, was ready to supply them with fuel.

As brave men hit the beaches and their landing craft ran the gauntlet of underwater obstacles and mines while under heavy enemy fire, behind them the supply operation was already running on a barely imaginable scale. Everything the invading army needed, vehicles, guns, tanks and ammunition had to be transported in vast quantities, as did food and most importantly, fuel.

The vehicles of the Allied invasion force required an enormous amount of fuel. Shipping it across the English Channel in fuel tankers was considered far too risky because of the danger of torpedoes from German submarines, aerial attack and minefields. Operation PLUTO was developed to pump petrol from England to France via specially constructed metal pipelines laid along the sea bed.

To protect D-Day the Allies had put into place major deception plans, including an entirely fictitious US Army Group, led by General Patton and stationed in Kent, to persuade the Germans that the landing would be made at the Pas-de-Calais. The Germans had assumed that if the Allies did choose Normandy or Brittany, they would have to attack a port such as Cherbourg to disembark safely and land supplies, including fuel.

Instead the Allies took their own floating harbour called Mulberry with them, and intended to keep the invasion force refuelled as it moved inland with PLUTO, a pipeline under the ocean, pumping fuel across the Channel from England.

It was to become the most audacious and complex engineering project ever attempted.

## *2.2.1 THE STORY OF PLUTO*

The planning required to supply any invading army is an immense task. It naturally involves having to provide ammunition, guns, vehicles, food and stores into a combat area in a constant stream. But the question of how to supply fuel for vehicles is potentially the most critical, for without it, and in never-failing supply, no army can move in a modern war. If that army has to be transported over sea and then has to attack an enemy that is dug in with heavy guns in fortified elevated positions, then the problem is almost insurmountable.

It had been recognised as early as 1942 that any successful invasion of mainland Europe would require, above all else, a huge, constant and reliable supply of fuel. However, that invasion force would have to rely on fuel tankers sailing across the English Channel to moor in French Ports which would have been extremely vulnerable to air, submarine or sea mine attack.

The first person to broach the idea of an underwater fuel route by pipeline was Lord Mountbatten. Mountbatten was a firm favourite with Winston Churchill and on 27th October 1941 he replaced Roger Keyes as Chief of Combined Operations and was promoted commodore. In 1942, Mountbatten instructed the Petroleum Warfare Department to explore the possibility of laying a pipeline across the English Channel. Nothing like this had been tried before in any theatre of war, or indeed anywhere.

The English Channel has exceptionally strong currents, amplified tides up to 6m and varying depths up to 120m (390 feet). It was known that the German Army had fortified the coastline with mines and other underwater obstacles. A pipeline was considered impossible, and the Petroleum Warfare Division didn't take the matter any further.

However, Clifford Hartley, Chief Engineer at the Anglo-Iranian oil company, was convinced the plan could work. He was also inspired by an Iranian pipeline that could transport liquid petroleum at high speed. Across the Atlantic, Hartley also saw how Shell had already developed technology to transport oil and other liquids across huge distances in the U.S. Also by 1941 the British had started to lay overland pipelines across the country forming the Government Pipeline and Storage System (GPSS), providing a secure fuel distribution network for the United Kingdom.

*Clifford Hartley - widely known as the father of PLUTO.*

This integrated network of 990 miles of pipelines was needed to move fuel from terminals (mainly Liverpool and Bristol) located in safer areas of the United Kingdom, across the UK. The GPSS network ended up being able to supply fuel to all of the airfields used by British and American bombers that carried out raids over Germany and occupied Europe. Post WW2 the GPSS pipeline was extended and developed to over 1,500 miles and today it connects all UK based refineries and major fuel processing depots as well as all major civilian airports - including Heathrow, Gatwick, Stansted and Manchester.

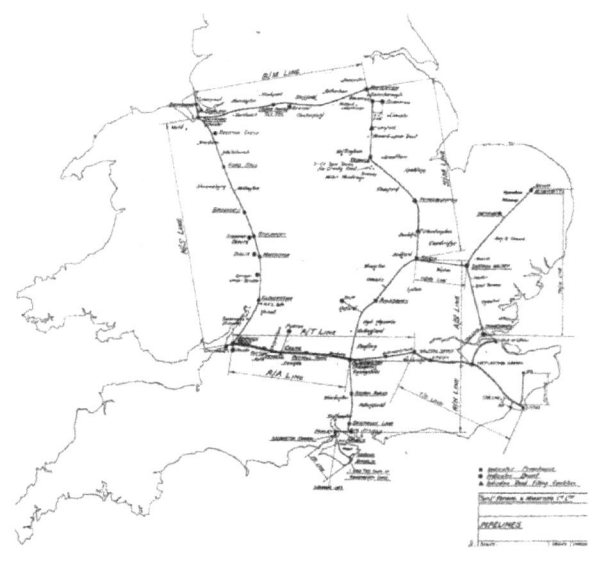

*Today the route of the GPSS system across the country has its own yellow topped marker poles, warning people not to dig or interfere with the pipeline below. In places, the original WW2 markers still sit alongside.*

But to lay high pressure pipelines under the ocean had never been tried. The British Army tested Hartley's lead pipeline, which was just three inches in diameter and found that it was strong enough to resist a bomb explosion underwater. Bernard Ellis also led British engineers from the BURMAH Oil Company in the design of a second pipeline, this time made of steel. In essence, with two types of pipe now available, PLUTO, the Pipeline Under the Ocean, was born.

The GPSS and PLUTO projects were carried out with the utmost secrecy; the pipeline channels across land were dug at night out of sight of enemy reconnaissance aircraft. The pumping stations and their power supplies also had to be disguised and were often hidden in plain sight under existing anonymous structures and even bomb damaged buildings.

In addition, it was realised that the fuel would need to be stored in large fuel tanks in the south of England in preparation for the invasion. All these would have been high risk targets for the Luftwaffe, so again disguise and camouflage were critical.

Although there were many moving parts and huge amounts of building and excavations works to be competed, the British government managed to keep the project hidden from the Nazis for nearly the entirety of the war. Anyone who transported materials to the various PLUTO facilities asked for 'Captain Jones' before they could continue.

The military planners for the Invasion of Normandy established a complete section devoted entirely to the supply of petrol, oil and lubricants (P.O.L). On D-Day and for many days afterwards, it was known that all P.O.L. would have to be taken over with the invading army 'packed' in jerry-cans.

A jerrycan (also known as a 'jerry' or 'jerrican') is a robust liquid container made from pressed steel. It was designed in Germany in the 1930s for military use to hold 20 litres (4.4 imp gal;) of fuel. The history of the jerrycan is notable because the robust and simple German design (hence Jerry) was reverse engineered and subsequently copied, with minor modifications, by the Allies during the Second World War in preference to their own containers which required tools and funnels to use.

For the D-Day invasion eleven million cans were made and fourteen thousand tons of petrol was preloaded. These would be landed with all other stores as part of the first assault, and with the ensuing battle these would necessarily have to be reused for some time. But sending the jerrycans back to England to be refilled and then returned was an appalling waste of time and cargo space. It quickly became clear that it was essential that the cans should be refilled from some form of 'bulk' storage as soon as possible on the French side.

Hence the ability to supply and receive bulk petrol close behind the invasion front was a major problem for the D-Day planners. The estimated requirement for petrol in the early days of the invasion was of the order of 1,000 tons a day for motor transport alone. For aviation a daily 700 tons was thought to be a moderate estimate. A Sherman tank had a fuel tank of just 137 (imperial) gallons (625 litres) which gave it a range of about 100 miles. General Patton's 3rd armoured division alone had 2,451 tanks and tank destroyers plus trucks and Jeeps. The planners were faced with a huge dilemma. No fuel meant no advance, and the very real possibility of counter attack, and the invasion force being pushed back into the sea.

*Jerricans stacked waiting for refilling from storage tanks. France, 1944.*

As part of Operation PLUTO the Army engineers had to develop schemes to erect, in a fantastically short space of time, large cylindrical storage tanks both in England and in France. They also had to get all the pumps and land pipelines ready. Due to the ferocity of battle it was expected that no French ports would be able to receive tankers for some weeks. However it was thought that Port-en-Bessin, a tiny port on the Normandy front, might be able to accommodate two small 400-ton tankers if the Allies were lucky, the idea being the ships could anchor and then fuel could be pumped ashore. 'Ship-to-shore' pipelines had been in use in various parts of the world for many years. Combined Operations, with characteristic energy, in conjunction with the Royal Engineers and Royal Army Service Corps, were soon experimenting with a floating pipe system, known as PLUTO MINOR at Ryde on the Isle of Wight. The team quickly developed a system that would allow tankers to pump ashore through pipes run out from the open beach, even though they would be exposed both to weather and enemy action. The ship-to-shore system would be codenamed TOMBOLA.

But a pipeline under the Ocean, once run, would be largely free of enemy interference and quite independent of the weather. But it would have to be small in diameter so several would be required to carry the volumes needed and the shore ends would need to come ashore into a specially prepared terminal. The experiments carried out by the Petroleum

Warfare Department Oil Engineers, working with the naval side of the Combined Operations and the Post Office had produced very promising results. By the early part of 1943, matters had progressed so far that it was decided that three different pipe systems would be required for PLUTO. But the manufacturing task was huge. It would require over 500 miles of pipeline, and all three types had to be developed from scratch.

The main cross-channel undersea pipelines would be codenamed as HAIS cable - essentially a hollow armoured cable with a lead sheath, flexible but expensive, and HAMEL cable – a cheaper and stronger thin walled steel pipe.

Overall this flexible, but expensive lead lined HAIS pipe was the favourite and easiest to use, but during wartime there were immediate concerns over the supply of lead and the time available to manufacture this type of pipe in large quantity. (HAIS from the initials of Clifford **H**artley the inventor, **A**nglo/**I**ranian his employer, and **S**iemens the designers & manufacturers.)

Two senior engineers, Henry Alexander **Ham**mick, Chief Engineer of the Iraq Petroleum Company Oil and Bernard J. **El**lis, Chief Engineer of the Burmah Oil Company working on the project already had experience of laying 3-inch steel (HAMEL) pipelines. They recalled that these were also still flexible when laid in long lengths and could also be joined to HAIS pipes where greater flexibility was required at the shore manifolds. This was welcome news and a parallel project was set up to find a second solution for PLUTO using steel pipes.

All the codenames issued across the PLUTO project were essential for security and even use of the words 'pipe' and 'pipeline' was forbidden, all concerned with the project being encouraged to refer to *cables* rather than pipes. Both pipe systems had to be capable of laying down their pipes on the sea bed in a single, continuous, fast procedure. But no such system yet existed. The plan was that by D + 10 TOMBOLA pipes must be able to supply at least 900 tons of petrol a day, while by D + 18 the laying of eight PLUTO lines might be started, each of which when completed would contribute 250 tons per day. It was an extremely ambitious plan.

Work on developing the technology to enable pipelines to be laid under the ocean began as early as 1942, but constructing flexible yet pressure resistant pipes and leak-free couplings was a difficult and slow process. The first prototypes were tested in May 1942 across the River Medway and in June in deep water across the Firth of Clyde using vertical triple ram pumps.

After successful full-scale testing of a 51.5 kilometres (31.6 miles) HAIS pipe across the Bristol Channel between Swansea in Wales and Watermouth in North Devon, PLUTO planning became a reality. However, the Bristol Channel experimental line was only two inches in diameter, but it achieved a throughput of 38,000 Imperial gallons a day at a pressure of 750 psi. The system worked and actually supplied fuel to Cornwall and Devon through 1945. PLUTO was finally a reality. Or at least possible. Hartley in the meantime, had also developed a larger bore 3-inch pipe capable, he thought of delivering nearly three times the amount of fuel.

It was always thought that during the invasion of Europe the Allied forces would attack the Pas-de-Calais, the most obvious and shortest sea route from the English coast. But a new and bold plan was hatched, to launch a mass invasion against the Normandy coast. Now a key part of the overall plan was to convince the Germans that the invasion would still be against the Pas-de-Calais. The British Army even commissioned an architect to construct a three-mile-long fake dock at the English ferry port of Dover, just across the Channel from the French port of Calais. King George VI even inspected the dock, which was replete with false pipelines, storage tanks, a fire brigade and anti-aircraft guns.

However, this radical new invasion plan threw the PLUTO project into turmoil. The new locations increased distance had never been anticipated and brought considerable new problems for the PLUTO team. The engineers calculated that it should just be possible to pump fuel from the southern shores of the Isle of Wight to the Cherbourg peninsula. The team immediately started to plan to equip larger ships for the Cross-Channel pipe laying work and suitable small craft to handle the landing operations. Dungeness had already been selected for the Straits of Dover pumping station and construction work had already commenced there, but now all efforts switched to the Isle of Wight.

The team immediately switched their attention to Ventnor and its neighbourhood. But as they clambered over the rocky coastline from Dunnose Head to St. Catherine's Point they looked in vain for a suitable sandy cove to lay pipes and build pumping stations. Also the currents and the jagged reefs off shore would have made working there almost impossible.

So the PLUTO team were forced back to Sandown Bay even though this added another three miles to their already tight pumping margins. However, the bay area did provide suitable accommodation for personnel and pumps, a good beach and a relatively sheltered anchorage for the delicate landing operations. It was also protected by the guns from Bembridge Fort, Yaverland Fort and Culver Cliff above, monitored by the Bembridge radar station and covered by undersea loop detection and minefield loops systems.

However, for D-Day the Army continued to pin all their faith fundamentally on TOMBOLA, which was classified as 'vital' and given top priority. They welcomed PLUTO as a valuable backup, but the idea of running pipes right across the English Channel in war together with the many technical uncertainties involved, was regarded as so problematical that the planners dare not count on it.

It was also clear that the naval side of pipe laying operation would have to be fitted into the entire massive 'naval movement' plan and carried out by a naval force specially equipped and trained. This was not an ideal situation for the military planners who were already stretched to the limit, trying to mount the largest seaborne invasion in history, against a defending force that had, over the past four years, significantly fortified the French coastline.

The Admiralty decided to create FORCE PLUTO that would operate all pipeline schemes under the sea from high water to high water. It therefore had the tremendous responsibility of guaranteeing the 'bulk' supply of petrol to both invading armies until such time as adequate ports would be in working order and able to discharge large tankers. It was decided to man this force from the Merchant Navy Pool, with men serving under the White Ensign, drawn almost entirely from the Royal Naval Reserve and the Royal Navy Volunteer Reserve. When completed, it numbered just over a hundred officers, a thousand men and twenty-eight ships of various types.

The PLUTO Force was commanded by a highly experienced senior naval officer, Commander John Fenwick Hutchings. After a distinguished background in submarine warfare during WW1, he had retired in 1934, but was brought back to help develop coastal defences in 1940. In 1943 he was promoted to be Commander Naval Force PLUTO, tasked with managing all contacts outside the naval service. Primarily he was in direct and constant collaboration with the Petroleum Warfare Department, the Army and Post Office, cable and pipe manufacturers, cable (pipe) machinery engineers and numerous shipbuilding firms. All these had to be managed, organised and coordinated. The PLUTO force was completely self-contained with its own staff and fell within the overall Command of the Allied Naval Commander, Expeditionary Force (A.N.C.X.F.), Admiral Sir Bertram Ramsay.

The main training and store base for PLUTO was located at H.M.S. *Abatos*, commissioned on 21st September 1943 in the bombed out remains of the Vickers Super-Marine Works at Woolston, Southampton. The remarkable camouflage scheme already in place rendered the site exactly suited to PLUTO requirements. A secondary base, HMS *Abastor,* was established at Tilbury in Essex for PLUTO training.

*Pipeline installation across the Isle of Wight was a huge engineering effort - pipes were laid under roads and train tracks while both were still in operation!*

Once engineers began pipe production, the next challenge was how to move the miles of pipeline and secure it under the Channel. The Hartley HAIS pipeline was transported to one of several loading docks, then slowly coiled around 50-foot drums inside the tankers hold, which would then deploy it into the water. This was an already a well understood process for deploying undersea telegraph cables. *(See picture on page 257.)*

However the Bernard Ellis steel HAMEL design, although cheaper to produce, was too fragile to be wound up in the same fashion directly onto a boat, as the material would simply snap.

So the Petroleum Department worked with the Miscellaneous Weapons Department, nicknamed 'wheezers and dodgers', to create something unique to the PLUTO operation. This was a giant, floating spool codenamed 'CONUNDRUM.' (Taken from 'cone-ended-drum)'.

*One of these huge drums could store up to 90 miles of pipeline. A tugboat would then tow the huge floating spool, which would slowly spin and release the pipeline into the water where it could gently sink onto the seabed. The huge Conundrums were actually built at Tilbury, where hundred miles of steel pipe were welded together and then wound on to them.*

*Conundrum.*

Training proceeded at once, though practically all the special craft needed had yet to be found and converted. Fortunately, the original experimental work had provided H.M.S. *Holdfast*, which had laid the experimental pipeline in the winter of 1942 across the Bristol Channel, and a barge, H.M.S. *Persephone* had just completed her conversion with a large wheel for laying HAIS pipes.

**HMS Holdfast** *was built as the SS London for the Dundee, Perth and London Steamship Company for passenger trade on the east coast of the UK.*

*On 29th August 1937 she sank after a collision off the Humber and was subsequently raised and repaired. On 28th August 1939 she was first requisitioned by the Admiralty for use as an examination vessel and in October 1939 was renamed as HMS Holdfast.*

*It was returned to its owners in August 1940 and then re-requisitioned in January 1942. Gutted and refitted by Green, Siley and Weir Ltd.; Two 30 feet diameter cable tanks were fitted, each capable of holding 15 statute miles of 2 inch HAIS pipeline. The cable machinery was loaned by the GPO and installed by Johnson and Phillips.*

*Following a series of tests, it was finally decided to carry out a full trial with the laying of a pipeline across the Bristol Channel between Swansea and Ilfracombe, a distance of 31.6 miles. In charge of HMS Holdfast was Captain Treby Heale RNR. Laying commenced on 29th December 1942 from Swansea and was successfully completed at a laying rate of 5 knots.*

*For the actual laying of PLUTO across the English Channel, HMS Holdfast was under the command of Commander Bicker Carten RN, and Captain Treby Heale took command of HMS Latimer deploying 3 inch HAIS pipes.*

It was now decided at high level that the first PLUTO lines would be laid across the Channel between the Isle of Wight and Cherbourg; so a pumping terminal (BAMBI) would be installed at Sandown Bay.

In France, Port-en-Bessin, which lies midway between the British and American invasion beaches would become the fuel port for both armies in the early stages of the invasion. It was not considered practicable, however, to bring the PLUTO undersea lines to that port on account of the considerable extra distance, so the PLUTO team would have to wait for the capture of Cherbourg before laying any PLUTO pipes across the channel. Immediately after D-Day, if all went to plan on the beaches, Force PLUTO would concentrate all efforts into hauling out the TOMBOLA pipes at Port-en-Bessin and providing large storage tanks near the coast.

To feed the PLUTO idea, the GPSS pipelines were extended down to Dungeness in Kent and across Southampton Water and then across the Solent to Thorness on the Isle of Wight. The ambitious plan was then to establish pumping stations on the Island and in Kent and then lay pipeline across the Channel to Cherbourg and Calais. This revolutionary design would supply fuel to the Allied armies on land, sea and in the air when they invaded Northern Europe.

For the undersea part of the PLUTO project it was necessary to link the Isle of Wight with the mainland. This was done by laying PLUTO pipelines across the Solent from Stone Point to Gurnard Head, within the 'prohibited anchorage' area. The topographical and tidal conditions here bear a very close resemblance to those existing in the Straits of Dover between Dungeness and Boulogne. But the work was made more exacting as the PLUTO team were firmly told that they must not on any account use Thorness beach, just to the westward of Gurnard Head. They also must also, on no account ever risk fouling either telegraph or power cables. Anyone who knows that area will realise that these restrictions narrowed down the limits in which PLUTO lines could be laid to very small dimensions. However, these restrictions and obstacles immediately demanded a new level of training for the team and forced them to rapidly develop new techniques.

Twenty-one pipes (Codenamed SOLO) were laid across the Solent in 1943. By 1944 twelve remained fully operational and were used to supply the pipeline across the Island to fill the storage tank hidden in Hungerberry woods above Shanklin. In his book, *Flame over Britain*, Sir Donald Banks, who was Director-General of the Petroleum Warfare Department during the War, provided some detail about the Hungerberry tank, codenamed TOTO.

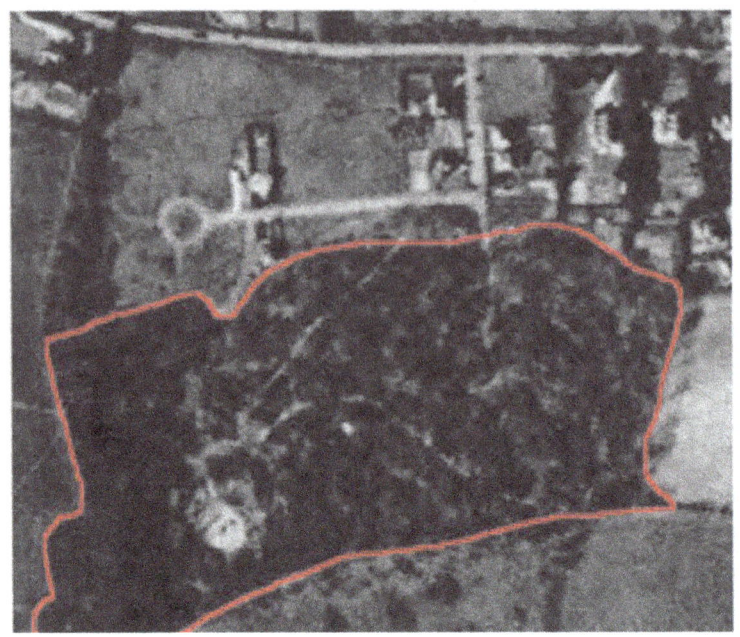

'At Shanklin a large 620,000-gallon tank (called 'TOTO') was erected in a small wood on the hill, the work being carried out entirely under the umbrella of camouflage netting of nearly an acre in extent. From there the [petrol] was gravity-fed down forking lines to two batteries of pumps installed respectively at Sandown and Shanklin, about three miles from each other.'

*Post-war aerial photograph of Hungerberry Wood. The bright circle (bottom left) is the scar left by the 60 foot wide and 30 foot high TOTO tank that was completely removed, (along with its camouflage) post-war.*

*A still image taken from rare film footage, shows the actual Hungerberry tank.*

The TOTO tank was located on the south side of Victoria Road, Shanklin in Hungerberry Copse. This was a completely covered area, where strictly enforced camouflage was maintained to hide the vulnerable petroleum tank. The whole area was covered by stretched fence wire on which steel wool had been laid. That area was then spray painted seasonally with greens and brown, like gorse. All the road turn-ins were cemented and paved and no truck was allowed to cut a 'short' corner or curb that would show that heavy traffic was running over the nearby dirt. In camouflage photography, texture is all important and good discipline included not running over vegetation that would show tell-tale tracks that would be a dead giveaway to enemy photographic reconnaissance experts.

*One of the authors inspecting a fuel storage tank at Cremyll, near Plymouth, similar in construction to the Hungerberry tank - these were used for fueling ships on the Tamar river.*

The increasingly complex and costly PLUTO plan then determined that two separate and discrete pumping sites were required. This would protect the system against enemy action not only by dispersion and concealment, but also by duplication. If one set of pumps or its power lines were knocked out, another could be switched in to take its place.

All the PLUTO pipelines were linked to pumping stations on the English coast, housed in various inconspicuous buildings; including cottages, garages and a cafe, all camouflaged from the air by their very ordinariness.

*On the Isle of Wight, the town of Shanklin had been badly bombed earlier in the war and, amongst the ruins of a Victorian hotel (the Royal Spa) and adjacent buildings, a pump-station was constructed. It was well hidden by the debris which was carefully returned to its same position once the pump rooms had been built, but to ensure its security extra camouflage was also added.*

*On the Isle of Wight, and later at Cherbourg the senior engineer was 2nd Lieutenant Maurice G. J. Lickens, Royal Engineers.*

The Shanklin PLUTO pumping site consisted of eight reciprocating pumps numbered 19-24 and 26-27, and one centrifugal pump numbered 25. The main difference between centrifugal and reciprocating pump is that, in centrifugal pumps, the fluid is continuously accelerated by a set of impeller blades whereas, in reciprocating pumps, the periodic motion of a piston draws in and discharges fluid. Reciprocating pumps discharge fluid in pulses, with each pulse discharging a fixed volume of fluid.

At Sandown the PLUTO pumping station was mainly constructed in the disused Palmerston 'Granite' Fort that had been built between 1861 and September 1864. The fort had actually been sold off in 1930 and was in the process of being demolished to provide building rubble as war broke out. Concrete machine gun posts were erected by a new generation of coast defenders in 1940, and the thick casemates of the old Victorian gun emplacements made perfect shelters for the pumps.

The site at Sandown was allocated sixteen reciprocating pumps numbered 1-11 and 14-18 and two centrifugal pumps numbered 12 & 13. (18 in total) with eleven PLUTO pumps (and their engines) housed in the fort with three more located in the buildings in the fort complex behind. Today, in what has become the Isle of Wight zoo, the old steel mountings and concrete blocks are still present in some of the gun ports.

*Two pumps were housed next door to the fort. One was installed in the glass roofed golf club (now the tea room) and the other in the front of the large brick built building (now administration offices) next to the Grand Hotel. At that time, this carried a sign reading 'Brown's Ice Cream' [Factory]. Today, inside this building, the original roof mounted lifting rail (below) still survives as each pump weighed more than seven tons and were all installed by hand.*

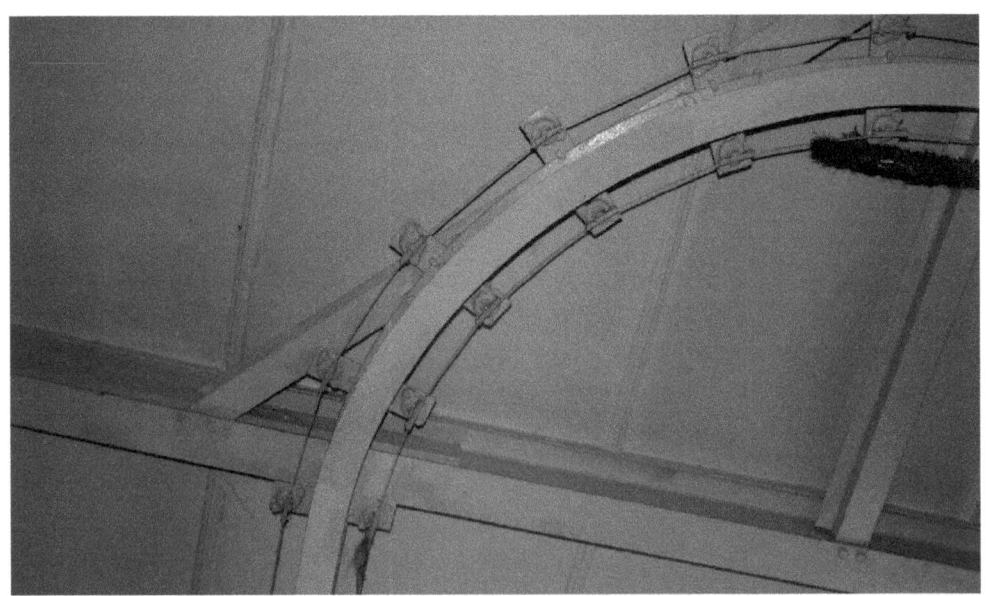

*PLUTO pump, restored and preserved in Sandown Zoo.*

One mystery that still surrounds the pumping stations at Sandown is the location of the PLUTO control room. The Shanklin control room location is also unknown, although one source refers to it being in the 'Osborne Annexe'. This could be in Osborne road on the cliff above, but more likely near Osborne steps by the cliff lift. Today a possibly wartime building survives (currently toilets)  this we think was the 'lift' pump room and may have been the control room.

Clifford Hartley's 1945 address to the Royal Society of Arts simply stated that the *'Main and group control rooms, with telephone communication between themselves [and] the pump houses, and to the opposite terminal [in France], were provided and had diagrams on their walls on which the control officers could indicate by disks on hooks the direction of flow of [petrol and] the pumps and lines, etc. in use at any time.'*

The only photographs we have are of the Dungeness DUMBO Control rooms. The precise role of the art deco Grand Hotel in the PLUTO story is also vague. For a long time, it has been assumed that the control centre for the BAMBI terminal was housed in the hotel's basement.

The Grand Hotel does have a basement, but no trace of PLUTO or WW2 installation, nor any original documentation has ever been found. Undoubtedly it was used for military accommodation during WW2 and for the PLUTO project, but it seems far more likely that the rear single storey part of the Brown's Ice cream factory, with a pump in the front room may have been the control room. Another possibility for the operations control room is the front of Brown's golf club 'pavilion', (now the cafe) although this may well have been the communications room. One recently found source (the Fire Alarm response documents for PLUTO!) states that there was a 'Fort Control' and a 'Golf Control', which again excludes the Grand Hotel (it has separate instructions for escape for PLUTO staff from this building). This document suggests that there were two control rooms, one in the fort for the fort pumps, the other in the Golf Club front building - otherwise surely it would have stated Ice Cream!

*Below is the only known photograph of Sandown Fort during wartime. Two PLUTO pipes can clearly be seen, along with road closing obstacles and security fencing around the fort. Bembridge Fort with its Chain Home Low radar dish can be seen in the far distance.*

*Dungeness Control Room*

The original planning for the PLUTO system had been that at full capacity the lines running from the Isle of Wight to Cherbourg (Codenamed BAMBI) would provide 3,300 tons of fuel per day. The erection of the pump houses was due for completion by January 1944. In a progress meeting on 15th February 1944, it was decided to stop work on the Shanklin site to concentrate on the Sandown site, due to the slow delivery of materials and lack of adequate workforce. It was expected that the Shanklin site would be completed by 15th April 1944 but this overran until the middle of May when most of the work at Shanklin had been completed. By the beginning of July, the pumps at Shanklin had been water tested and by 5th August the pumps had been spirit tested and certified ready for operation.

Caterpillar diesel engines (40 of the 60 total were shipped over from the USA) were installed to provide power for the reciprocating pumps, while the centrifugal pumps were powered by electric motors, connected to a high voltage electrical supply. This had been specially routed across the Island by the Isle of Wight Electric Light and Power Company on the river Medina to an adjacent transformer house. However, from recently found documentation it appears that the high voltage supply for the Sandown site may never have been completed. It was run to Bembridge Fort in early 1940 to power the radar station(s) installed there, but no record of it being extended down to Sandown sea front and the Granite Fort can be found.

Each PLUTO pump had a capacity of 36,000 gallons (163,659 litres) per day and it is stated that each pumping station could deliver fuel at a pressure up to 1,500 psi (105.46kg/cm$^2$). However, after much research the authors are still unclear how this was achieved. One clue is a statement that 'to achieve the necessary pressure, the set-up involved centrifugal pumps receiving fuel on the suction side from an 8-inch main pipeline, before delivering it to a number of reciprocating pumps arranged in series, to provide the necessary pressure needed to pump the fuel the long distance to Cherbourg.' John Sullivan states that 'the reciprocating pumps were connected in a serial sequence so one received the pressure from the last pump and added its pressure to it. Each one was set to add 100-pounds pressure. Hence, h states, that set sequentially, 16 pumps could deliver fuel at 1,500 psi discharging into four high-pressure bus mains running parallel to the coast. From these bus mains, which have isolating and balancing valves, the lead-offs to the sea pipe lines were taken. At the start of each sea line there were twin high-pressure filters fitted to prevent dirt or sediment becoming lodged in the pipe.' But there are some quite serious technical laws with this, not least of which is that the pumps couldn't have functioned in this manned. If there are any pumping engineers who would be able to try and work out a possible solution please contact the authors!

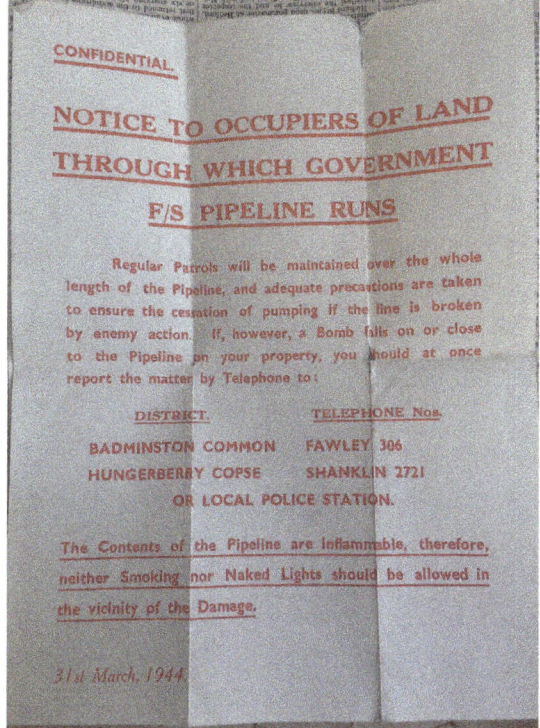

To prevent the Germans from becoming suspicious about all this work if they were spotted during reconnaissance flights, a common story was given out at the time (still believed today by some people) that the pumps were just there to water the golf course. The general cover story for the 14-mile pipeline across the Island was that it was only drainage, irrigation or general water pipe works.

*However, this subterfuge was somewhat lost when a letter was issued in March 1944 to the local farm owners and police stations, stating that the contents of the pipe were flammable.*

Incredibly, against all the odds, and in an almost unbelievable timescale, PLUTO was now possible. A complete Outline Plan of the operations of Force PLUTO, including the laying of TOMBOLA pipes behind the assault, was now made by the Senior Naval Officer (S.N.O.) PLUTO and submitted to A.N.C.X.F. and to the American High Command.

This Plan was approved and PLUTO became a small part of the most stupendous example of complex naval movement organisation the world had ever seen.

The **PLUTO FORCE** was organised in six divisions, each having a special function.

**The 1st Division,** consisting of barges and boats, was to operate with the PLUTO ships in Sandown Bay.

**The 2nd Division,** with trawlers, barges and boats, was detailed to carry out the TOMBOLA operation at Port-en-Bessin.

**The 3rd Division,** with a small cable ship, barges and boats would be at Cherbourg when the PLUTO lines were being laid.

The big cable ships formed the **4th Division** and the Conundrums, with their tugs, the **5th Division.** The pipeline from the Isle of Wight would be Codenamed BAMBI.

A flotilla of six motor fishing vessels forming the **6th Division** was intended to work with the tankers at Port-en-Bessin.

During operations S.N.O. PLUTO had H.M. Corvette *Campanula* as his flagship and later, for the Dover Strait operations he used Motor Launch ML197. On D + 1 Port-en-Bessin was captured and the 2nd Division sailed at once with the S.N.O. in the *Campanula* to commence TOMBOLA operations. The foreshore at Port-en-Bessin was not particularly suitable for this operation, being very rocky, and the infamous gale of D + 13 caused some delay. However, in the ensuing weeks four pipes were built and hauled ashore; two at Port-en-Bessin for the British and two at St. Honorine des Pertes for the Americans. Three more TOMBOLOA pipes were hauled at the special request of the Americans at Fox Red Beach in the OMAHA area.

Fuel for the Allied armies on D-Day could now be delivered by dual-purpose tankers. In addition, in anticipation of bulk delivery, concrete storage barges, each capable of holding 180 tonnes of fuel, were towed across the Channel.

On 25th June, the first ship to shore TOMBOLA pipeline was completed and fuel successfully flowed through it on 3rd July.

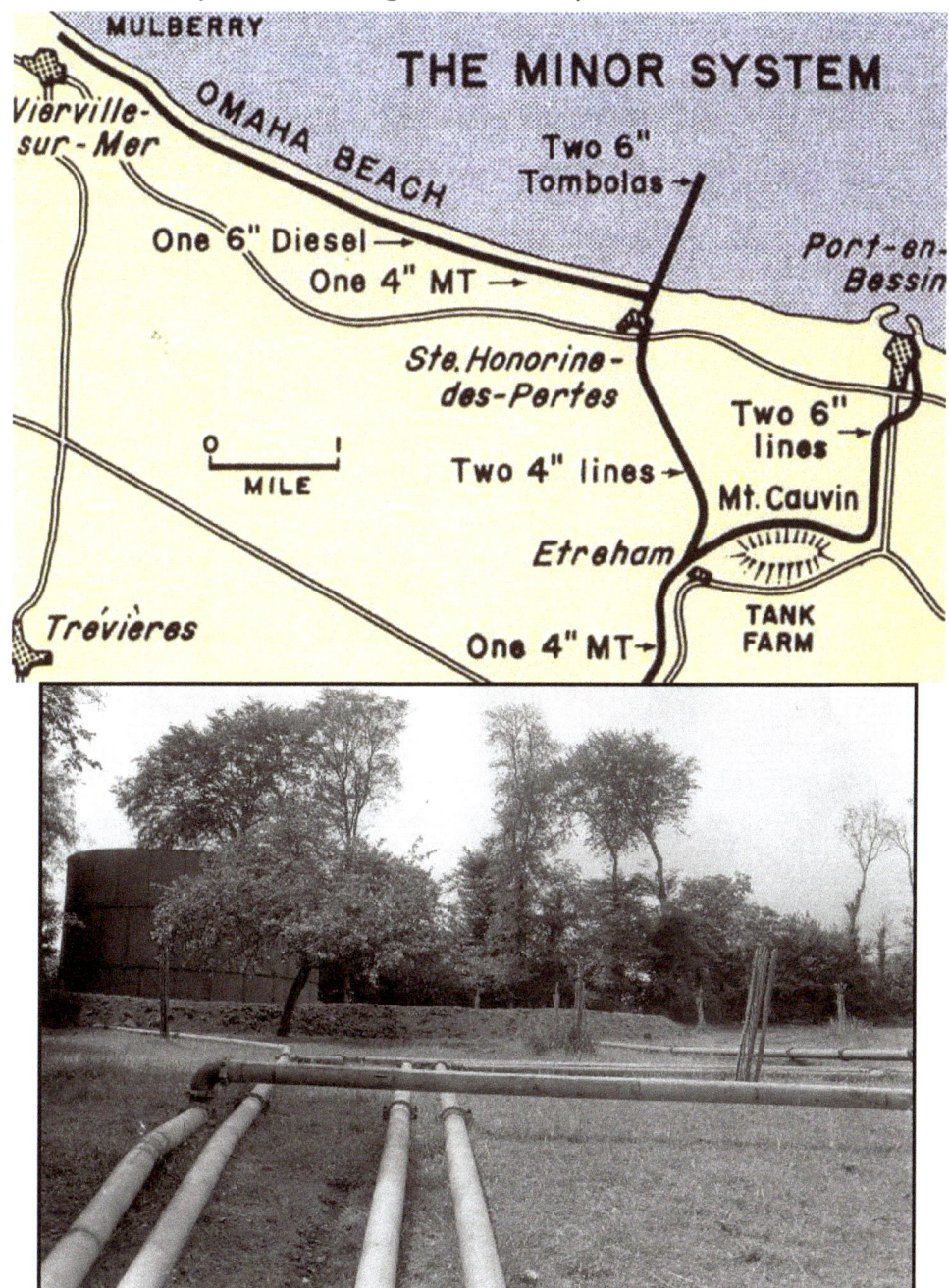

TOMBOLA pipelines and storage tanks.

The pipelines from Port-en-Bessin would be joined by two 6 inch TOMBOLA pipelines bought ashore at Saint Honorine-des-Pertes, 5km from Port-en-Bassin.

All this time the Americans were fighting a brutal battle for Cherbourg, and the port fell about D + 20. The clearing of obstructions and mines took a considerable time, so that the PLUTO Force from the Isle of Wight could not start operations until D + 36.

The first PLUTO pipeline had been scheduled to be laid on 24th June from Cherbourg harbour, but this of course relied on capturing the actual harbour intact in just 8 days. It was not taken until 27th June. The Germans, not surprisingly, had also systematically demolished all the port facilities. Over a month was now spent in debate as to whether the pipeline terminal should be located outside or inside the breakwaters, until it was agreed with the American Command to make Nacqueville Bay, just west of Cherbourg, the receiving terminal for PLUTO supplies of petrol. This bay was still heavily mined and obstructed, but it was open to the north.

By now PLUTO was now so far behind programme that some naval authorities suggested that it should be cancelled, but the situation was becoming critical as TOMBOLA was proving to be only a limited success as the pipelines kept breaking on the rocky foreshore.

For tactical reasons it was decided to bring the loaded pipe laying vessels over to Cherbourg and lay the PLUTO pipes from France to England. The laying ship was escorted by five corvettes, two anti-submarine (A/S) patrol boats and full air cover was provided by fighter aircraft. The S.N.O. in H.M.S *Campanula* formed part of the screen and was in overall command. The laying ship could only proceed at about five knots and could not manoeuvre to avoid attack; also on no account could she stop as when laying both HAIS and HAMEL pipes it was an imperative rule: -

> 'One must never go back on a pipe but maintain it always in tension in the direction of the lay'. Failure to carry out this rule inevitably results in kinks and trouble, but to carry this out with the immensely clumsy types of vessel in use was a matter of supreme difficulty requiring much skill.'

Once the convoy was underway it was also essentially helpless. All would depend on the escorts being able to protect the pipe laying operations from enemy action.

A sea channel two miles wide had to be swept and a flotilla of minesweepers preceded the laying vessel; but to keep within a margin of one mile at a speed of advance of five knots in a three and a half knot cross tide takes some doing. The use of radar position-finding was essential to this achievement but required constant checking to meet the varying conditions of tide and speed.

The first cable run was made by H.M.S. *Latimer* on 12th August 1944. It was a perfect day; she started just before dawn with overhead cover from fighter aircraft and an escort of naval patrol vessels and entered Sandown Bay, securing to the buoy there off Shanklin pier within five minutes of her expected time of arrival. The run had taken about ten hours.

The escorting vessels also anchored in the bay. The team were preparing to bring the end of the first cable ashore when they saw signals hoisted by one of the patrol ships asking for instructions. It was not until later that the full calamity was revealed. The escort had dropped her anchor hook and picked up the precious cable in a hopeless snarl. The whole pipeline was wrecked. She certainly needed instructions, which were forthcoming in terse naval language. But the first PLUTO pipe laying operation with it huge outlay of time, materials, man power and risk came to naught.

Two days later H.M.S. *Sancroft* set out for Nacqueville and laid another line successfully across from Cherbourg. Again disaster struck when attempting the link up to the English shore; H.M.S *Algerian* got the pipeline wrapped around her propeller, which immediately destroyed it beyond use. Over 150 miles of valuable pipeline was now scrap at the bottom of the sea.

> Sir Donald Banks wrote:
>
> 'The technique of cable laying had been mastered but we were not yet sufficiently versed in the practice of connecting the shore ends, nor in effecting repairs to the undersea leaks which were caused fairly close inshore through these faulty concluding operations.'

While the two ships returned round to the Thames to reload with HAIS pipe at specially arranged loading berths, the next step was to try to lay the steel HAMEL pipes. For the first time the great Conundrums were brought into use.

The HAMEL steel pipe process underwent similar teething troubles. It was a particularly severe test for this system, as the drums had to be very heavily loaded to carry the seventy miles of pipe necessary for this long span. The tides swept past the Cotentin Peninsula at breakneck speed and the rocky bottom was anything but favourable for the pipe. Originally planned to carry thirty miles of pipe, the new requirement of seventy-two miles for the Cherbourg run necessitated heavy overloading which submerged the drum up to its trunnions, with a draught of over four fathoms. (7.5m)

Seventeen layers of pipe were wrapped around this giant spindle which weighed over 1,600 tons. It was a huge weight to pull and the submergence increased the water resistance greatly. The most powerful tug available, H.M.S. *Bustler*, failed to move this massive weight at more than three knots, but as the Channel tides can run higher than this, the future for PLUTO looked bleak. Clearly more tugs were needed, but tugs of great power were in such demand for the Mulberry Harbour and the Invasion that it was said that Churchill himself kept the list of them in a locked drawer in his desk, and only allowed their use with his personal sanction. But the PLUTO team approached Sir Hastings Ismay and obtained another tug, H.M.S. *Marauder* for the team.

This new tug only raised the convoy speed to about four knots, but the team were willing to try. However, the first attempt on 27th August failed. Some ten tons of barnacles from Southampton Water had attached themselves to H.M.S. *Conundrum I* during the wait for the liberation of Cherbourg. The second and third attempts also proved abortive, the pipe breaking on the latter occasion about twenty-nine miles out from Cherbourg, probably owing to it fouling and cutting on the sharp flange of the drum. Nearly another 250 miles of pipeline was again a complete loss, as no repairs could be made at sea.

H.M.S. *Conundrum II,* however, had better luck and on 29th September 1944, made an excellent run in a fairly heavy sea and delivered the pipe without mishap. However, a serious issue soon became evident when the pipe ends had to be lifted out of the sea to be connected to the manifold on shore. On the approaches to Cherbourg beach, the heavy pipe had to be pulled up the beach, but engineers calculated that the power required was far beyond the limit of any winches available to them.

There was, however a most unlikely solution from an earlier age. Hutchings recalled a boyhood memory of watching two steam powered ploughing engines at work. A phone call to the Ministry of Agriculture and Fisheries resulted in six large engines being allocated to the PLUTO project. It appears that two engines went to the Isle of Wight, one each to Sandown (called SUE) and Thorness Bay. One went to Lepe at the entrance to Beaulieu River on the mainland opposite Thorness and one to the PLUTO training exercise area at Hengistbury Head near Bournemouth. Another one, a Fowler class BB1 (No 15220), built in 1918, was sent to France. This was given the name STEVE, and when it arrived in Cherbourg it amazed both the French and Americans, who had never seen anything like it before in the war. It was both archaic and effective. The engine's modified hauling drum exerted a 14-ton pull to bring the pipes ashore and these engines became great favourites with the PLUTO teams. When finally stood down, STEVE was proudly decorated with a brass plate with its name and services inscribed.

*Rare photographs showing PLUTO pump installations with the typical arched section concrete roof. Location unknown, but likely Dungeness.*

*The Shanklin pumping site was heavily camouflaged in bombed out buildings.*

*The Boulogne Beach Pipes and Manifold connections.*

*Fowler Steam Traction engine of the type used to pull PLUTO pipes onto the beach.*

*The Cherbourg Beach Pumping Station.*

Once connected at both ends, the pipeline had to be tested. During laying the HAIS cable had to be filled with sea water in order to give necessary strengthening to the hollow core to enable it to withstand the strains of passing over the laying machinery and the pressure of the sea when it reached the bottom. At intervals in the Cross-Channel length thin metal discs were inserted to contain the water and these were devised to burst at a pressure of 400 psi.

When the laying operation was reported complete the pumps were coupled up and more water was pumped into the pipe from the home end. Anxious faces would gather round the pressure meter in the control room to watch the needle climb steadily to the bursting pressure of the first disc and a sigh of relieve would go up when it suddenly wobbled and fell back again. The first disc had blown satisfactorily. Successively one disc after another would be negotiated, the excitement growing as the last ones were reached.

Eventually, some 1 ¾ hours after the commencement of pumping the final one would go and a welcome telephone call from the other side would announce 'Line on flow'. Water was gushing out onto the beach. Petrol would then be substituted for water and it normally took about 24 hours before the first traces appeared across on the other side, when a halt was called while the pipe was connected into the tank system.

The residual water was strained off by the simple expedient of letting it settle to the bottom of the tanks and draining it from there, the fuel floating naturally on the top. As soon as the fuel flow in the pipes became pure enough for direct use the line was pronounced operational. During the process aircraft were sent out to spot any oily patches from possible leaks on the surface of the water.

Eventually one HAIS line was completed and water-tested on 18th September and on 22nd September was brought into commission. It was reported to have delivered 56,000 Imperial gallons of petrol at a pressure of 750 psi. The Quartermaster-General was in Scotland at the time and sent a telegram to the PLUTO team.

> 'Well done the King of the Underworld.'

On 29th September a single HAMEL pipe also became operational.

But it was to be a very short lived success.

Sir Donald Banks wrote:

> 'On 3rd October the elation was changed into funereal gloom when it had been decided to increase the pressure on the HAIS cable. It climbed steadily without incident to over 1,000 lb. per square inch when suddenly in the middle of the night the indicators went wild and dropped to next-to-nothing.'

The precious HAIS cable that had taken so much effort and so many failed attempts had failed.

Later, on the much shorter DUMBO crossing, pressures of 1,200 psi were achieved without any trouble at all. A little later the same night, the duty officer reported that the lone HAMEL also had failed, possibly breaking across a sharp edge on the sea bottom or simply that one of the many connections on the pipe had failed.

The whole PLUTO project was now non-operational, and would require new pipes to be laid. But there was another problem facing the Isle of Wight BAMBI team. By the time the HAIS flexible and the HAMEL steel pipeline to Cherbourg were finally ready to pump petrol, the Allied advance had been extraordinarily rapid, and supply problems for the armies far from their original bases in Normandy were assuming serious proportions.

The pipeline network on the Continent, starting at Port-en-Bessin, had been linked up with Cherbourg, where tankers direct from America were now discharging, and the line had been extended at a prodigious pace by the U.S. Engineers south of Paris. A spur ran to the Lower Seine crossings, but the quickest way to lengthen this spur to feed Field Marshal Montgomery's forces in the Low Countries was to work straight across the Straits of Dover.

In addition, between 1st August and 4th August 1944 seven divisions of Patton's Third Army had swept through Avranches and over the bridge at Pontaubault into Brittany. The Allies were able to advance freely through undefended territory and by 25th August all four Allied armies (First Canadian, Second British, First U.S., and Third U.S.) involved in the Normandy campaign were on the river Seine and the 4th Infantry Division, a lead element of Patton›s Third Army, arrived at the outskirts of Paris. Allowing the French 2nd Armoured Division to take the lead in the liberation of their capital, the division moved into the city.

Just five days later, Operation Overlord, the Allied invasion of Northern France, was declared over. PLUTO had not managed to pump a single litre, and was still a month away from operation.

But Patton was not done. He had his eyes set on Germany and continued to push his forces hard. As Third Army drove hard towards the French province of Lorraine, they finally outran their supply lines and on 31st August 1944, Patton's advance ground to a halt. Patton had assumed that he would be given priority for fuel supplies due to the success of his offensive, but was furious to learn that this was not the case. Eisenhower favoured a broad front approach and allocated more incoming supplies to Montgomery for his bold plan – Operation Market Garden, to be fought in the Netherlands. Despite their success in defeating German units all across France and driving further than any other force, the men of Third Army would have to wait for their chance to drill into the heart of Germany.

BAMBI from the Isle of Wight would not be ready to pump fuel for another month, until the 29th September. In reality this was a disaster and arguably a major failure in military planning. Any fuel from the Cherbourg terminal was now in the wrong place, and too far away from the battle that was rapidly heading west towards Germany and eventual victory.

The next step in the PLUTO story came when the heavily fortified port of Boulogne in northern France was finally taken by the Allies.

It fell after five days (September 17th -22nd) of heavy fighting during Operation WELLHIT and in a subsidiary operation the Allies captured the German long-range heavy artillery battery at Cap Gris Nez, which threatened the sea approaches to Boulogne.

With the failure of the BAMBI pipelines, all PLUTO efforts were switched to the Dungeness to Calais pipelines, and Hutchings set about laying pipelines across the Narrows.

*Commander John Fenwick Hutchings inspecting one of the PLUTO centrifugal pumps at the Dungeness Pumping Station for DUMBO.*

On 22th September 1944 the PLUTO route across the Straits of Dover, code name 'DUMBO' was ready and H.M.S. *Sancroft* made its first run on 10th October 1944. It was completed without incident, based on the experience gained from the BAMBI pipeline work. A good deal of difficulty was experienced in bringing in the shore end on the English side, as the shallow beach made it difficult for the Navy to deliver the cable to high-water mark – their appointed limit. Lieut. Colonel Danger's men of the Royal Army Service Corps, who were responsible for the manning of the pumps both at

DUMBO and BAMBI, stood for hours up to their waists in the very cold sea water, easing the heavy length ashore, and made a fine job of connecting it to the manifolds and ring main at the Dungeness pumping station.

As the fighting moved closer to Germany, 17 more lines (11 HAIS and 6 HAMEL) were laid from Dungeness to Ambleteuse near Boulogne in the Pas-de-Calais to shorten the supply route. Through these short 20 mile pipelines over 600,000 tons of petrol would be pumped.

Finally, but more than four months after D-Day, PLUTO was operational.

*HMS Sancroft.*

*27 miles of 3 inch HAIS pipe being paid out to the stern from the hold of HMS Sancroft.*

# Where PLUTO Crossed the Path

*Rambles with a Purpose*

*on the Isle of Wight*

SECTION 2 - THE PLUTO STORY

CHAPTER 2.3

## *PLUTO ACRONYMS and CODEWORDS*

Without delving too much into the overall PLUTO project, which is best left to Adrian Searle's book, I have always found the choice of the codenames used intriguing.

In recent years the procurement part of the Ministry of Defence have had an alphabetical list of random words and names to draw on for military project code names and as a new project is launched the next name on the list is given to it. It is unclear if this was always the practice during WW2.

Of course the actual name PLUTO has various origins – originally it was designated the **Pipe Line Underwater Transportation of Oil.** Today it is now more usually taken as the **PipeLine Under The Ocean.**

PLUTO, the main project concept, would certainly have preceded SOLO (The codename for the pipelines running from Lepe to Thorness across the Solent and TOTO (The Hungerberry Storage tank) and that, with the 'S' and the 'T', would fit the chronological practice even though SOLO would appear to be, quite rightly, connected with the Solent!

Despite the recognised acronym for PLUTO (and the apparently partial one for SOLO) there appear to be other subtleties involved and it would be interesting to know if they were intended. PLUTO of course was the Lord of the Underworld (aka. Hades) which in itself was very fitting, but PLUTO was of course also the dog of Disney cartoons and TOTO was Dorothy's dog in The Wizard of Oz!

The codename for the Shanklin and Sandown pumping stations was BAMBI, and later the pumping station at Dungeness were called DUMBO, also following the Disney theme. One has to wonder if there was a major American influence at work here!

However, these names do fit the recent MOD practice, one alphabetical list having run its course, they would then have embarked upon another, hence the 'B' and the 'D'. But of course, the 'DU' from DUMBO fits in with Dungeness in the same way that 'SOL' from SOLO fits in with the Solent. So if there was a standard practice at the time it certainly appears to have been conveniently manipulated.

The codenames do get even more complex when Sandown and Shanklin were named BAMBI NEAR and Cherbourg became BAMBI FAR. Equally Calais became DUMBO FAR and Dungeness DUMBO NEAR.

To add to this growing complexity is WATSON. The initial PLUTO pipes that were laid to Cherbourg were given the code name WATSON. The first Bournemouth Bay test pipelines were given the codename TWEEDLEDUM, and the pipeline system designed to connect on shore storage tanks to off shore fuel tankers was called TOMBOLA.

Mythological tradition also seems to have asserted itself at some point, and PERSEPHONE, the name of PLUTO's six-monthly wife, was given to Hopper Barge W.24 when she was converted with an enormous wheel in her middle to lay HAIS pipeline, whilst the PLUTO Base at Woolston, Southampton, was named ABATOS, a polite synonym for the Lord of the Underworld's dwelling place.

Whichever department was the equivalent of MOD in assigning codenames during the war, it was certainly playing word-games!

Dare I suggest that its apparent manipulations of any intended random principle could well have led to the enemy, if they had got wind of it, putting two and two together! (Or should that be TO and TO?).

*JF with TW. June 2019.*

# Where PLUTO Crossed the Path

*Rambles with a Purpose*

*on the Isle of Wight*

SECTION 2 - THE PLUTO STORY

CHAPTER 2.4

## *TYPES OF PLUTO PIPES*

*Sainte-Honorine-Des-Pertes - Tombola pipeline across the beach*

**TOMBOLA** pipe. This was what is called A.P.I. casing (steel), of six inches' internal diameter. Short lengths were jointed together by screwed sleeves into 400-foot sections. These were flanged together successively as they were hauled out to sea. On the sea end was a wooden sledge to ease it over the bottom, or, as used at Port-en-Bessin, over rock, a special split buoy with a spherical end. A heavy flexible pipe was connected to the sea end when that was fixed, the free end of the flexible pipe being sealed and buoyed. Moorings were disposed about the pipe end so that a tanker lying to her own anchors ahead could have her stern held firmly over the pipe end. She would haul up the flexible pipe and connect it to her own system and refill, or pump her valuable cargo ashore.

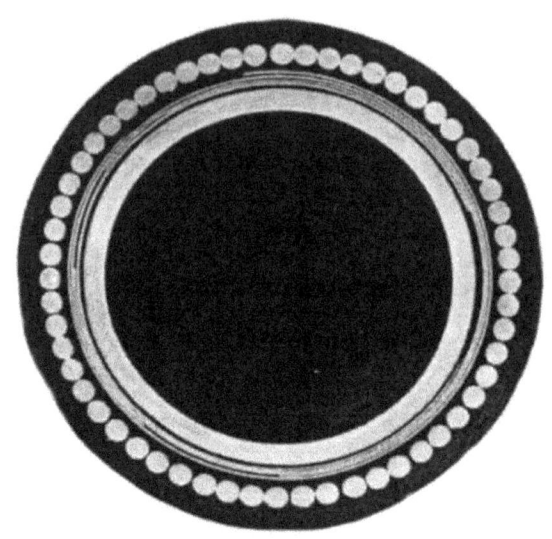

Cross section of 3-inch HAIS Cable

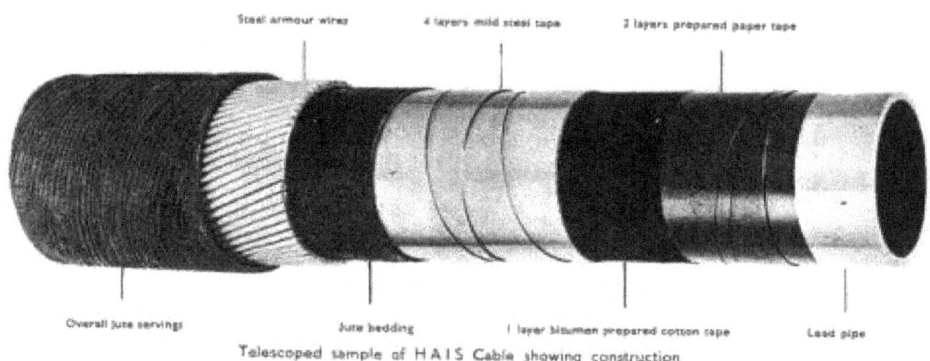

Telescoped sample of HAIS Cable showing construction

**HAIS** (Cable) Pipe. This is a lead pipe of three-inch internal diameter. It was made in the same way as an electric cable but had four layers of steel tape wound on to give it strength to stand internal pressure, and was armoured to give it tensile strength. Made in 40-mile lengths, each length was finished at each end by a special joint flange and then sealed with a copper disc. Each length was then filled with water and pumped up to a pressure of 180 psi (just like a tyre), that assisted in handling when laying and prevented kinks. These discs burst at 300 psi (some sources state 400 psi) during the pipe commissioning process. The HAIS cable was laid from a cable ship like heavy electric cable, but it was much more delicate and tricky to handle.

**HAMEL** Pipe. This was a steel pipe of three-inch internal diameter. It was manufactured in short lengths which in turn were welded together into 4,000-foot lengths at Tilbury and stored on long racks. Then it was wound onto the huge Conundrums, 40 feet in diameter and 60 feet long between 6-foot flanges, held in arms, floating, at the end of special jetties and turned by chains on the flanges.

Each 4,000-foot length was welded to its successor and it was slowly wound until thirty or seventy (perhaps 80) miles was on the drum. HAMEL pipe had about 5 per cent less throughput than HAIS owing to the rougher surface of the pipe. It was laid by towing the Conundrum with a powerful tug and slowly unreeling the pipe as it went along. Another tug astern assisted the steering and control the drum, and was useful in handling the pipe at the beginning and end of a run.

There was an alternative pipe to the **TOMBOLA** pipe, should it have been found impossible to haul the pipes out of the sea. This was a ten-inch floating steel pipe two or three thousand feet long which would be towed into position and sunk; this very difficult operation was practised at Ryde. A number of these pipes were made up at Exmouth and moored ready on the sands there. They had a code name **AMETHEA**. As things turned out they were not required.

One PLUTO manufacturer was Callender's (now BICC Cables) based in Erith, Kent, on the south bank of the River Thames. Once production processes had been established, 250 miles of HAIS cable in a year was produced working round the clock seven days a week. The material required to do this was: 6,843 tons of lead, 2,500 tons of steel tape, 4,250 tons of galvanized steel wire, 275,000 yards of cotton cloth, 540 tons of jute and 1,100 tons of petroleum compound. W.T Henley in Gravesend consumed 8,000 tons of lead and 5,600 tons of steel wire. The pipe was usually manufactured in lengths of 40 miles weighing 2,000 tons. The weight of the cable, when filled with sea water was 67 tons per nautical mile.

# Where PLUTO Crossed the Path

*Rambles with a Purpose*

*on the Isle of Wight*

SECTION 2 - THE PLUTO STORY

CHAPTER 2.5

## *BAMBI - WAS IT A PIPELINE TOO FAR?*

*There are many questions still to be answered about the PLUTO Project and especially PLUTO on the Isle of Wight.*

The wartime PLUTO project was a huge engineering effort with an almost impossible timescale. It faced and conquered vast new technical, engineering and logistic challenges.

Because of this the army always considered the undersea PLUTO pipelines to be something of a Hail Mary, possibly able to provide fuel to the invading army, if all else failed.

By 1944 the nation had already been at war for four years and was critically short of resources and manufacturing capacity. It was also a nation preparing for the greatest amphibious landing in history. Yet still the huge PLUTO project was planned, designed, built, installed and then made operational. But when it was eventually deployed, under extreme combat conditions, it is clear that the technology and techniques developed in such a short time were at the very best immature, and at worst, not fit for purpose.

The PLUTO project had absorbed a huge logistic effort and used up massive quantities of materials and manufacturing resources. It also swallowed large amounts of manpower both military, reserve and civilian and required continual and long term retasking of vital ships and military equipment, long before, and long after D-Day.

**As for what actually happened on the Isle of Wight?**

The first unknown is the actual number of pipes that were laid across the Channel to Cherbourg from the Isle of Wight as part of BAMBI. We think, from all the sources at least 5 pipes were laid, 3 HAIS (although it is likely another HAIS was laid as the ships were working in pairs - so 4 in total) and 2 HAMEL. But only one of each pipe type was laid successfully, the other three or four, nearly 300 miles of pipe, were abandoned on the sea bed.

It is clear that at best, these two BAMBI pipes only survived in use for perhaps two days before they failed.

On 3rd October 1944 the one report we have states that when the pressure was increased to increase the amount of fuel being pumped, both pipes failed within hours of each other. Then every pumping station, pump, generator, storage facility and building involved with PLUTO on the Isle of Wight was abandoned and the team simply went to Kent.

There is no clear explanation of why the two 'working' pipes stopped. We only have Donald Banks' word that in those two days any fuel moved at all. Original sources are at best conflicted and the figure of 3,300 tons of fuel being pumped (in total) over two days of operation is suspiciously identical to the planned daily total for PLUTO quoted elsewhere (for DUMBO).

A team of American engineers involved in Cherbourg working with the TOMBOLA system were dismissive of even this fact. Indeed, is there a risk that the BAMBI system never managed to deliver any fuel, and were Bank's words written just post-war (still under war-time censorship rules) in fact a face saving exercise?

Today it is clear that these pipelines under the ocean were never going to be strong enough to withstand the constant pounding of tides, currents and the rough seabed that would have continually moved the pipelines around. This was probably the cause of most pipe failures, especially at DUMBO. Add in the large drops from sea level on both side of the channel – where the pipe line has to plunge down from the coast, span the various contours in the English Channel and then climb up to the French coast. The sea bed of the English Channel is for the most part smooth sand and mud that changes to gravel and pebbles in places. But there is one area along the route to Cherbourg where an extensive rocky area lies before the deep channel in the middle. The pipeline would have to drop down across these rocks and would have been moved side to side across them. From Sandown the pipeline ran in water from 33 feet (10m) gradually descending to 99 feet (30m). The middle section of the mid channel had to drop from 254ft. (77m) and at its deepest 272 feet (83m) where the external pressure is approx. 9.4 Bar or 136 psi. The pipe then had a near vertical climb from 92 ft. (28m) to get the fuel to reach the beach manifolds on the French shore. This is a significant extra pumping load when compared with a horizontal land based pipeline such as GPSS.

The change in planning to ultimately invade the Normandy beaches over trebled the planned and designed route for PLUTO long after the project had been started. This fact alone effectively doomed the PLUTO link from the Isle of Wight to Cherbourg. But as ever the engineers persevered and Operation PLUTO with its pipeline systems still under development, had to make rapid adjustments to try to cope with this increased distance.

The American TOMBOLA team. documented in *Fuel to the Troops: A Memoir of the 698th Engineer Petroleum Distribution Company 1943-1945* by John G. Sullivan had been briefed on PLUTO as they were to do the tie-in to the pipeline on the European side. (Interestingly, there is no record of them ever having doing so at Cherbourg for the undersea pipelines.) The PLUTO HAIS pipe designed to deliver petrol was 2⅞ inches inside diameter, and about 3⅛ inches outside diameter. One of their team, Bernard Crawford had a Petroleum Engineering degree and had made a rough estimate of the coefficient of friction of a 2 ⅞ inch line. He was convinced that the distance across the Channel was an impossibility. Crawford simply stated that you could not pump at any pressure to overcome the friction loss. He estimated that probably about 18 to 25 miles was the limit on a small pipe like that, at any amount of pressure. Crawford had apparently listened to the English Major explaining PLUTO and put up his hand.

He said: 'I'll stand over on the other side with a five-gallon bucket, and carry off every drop you can pump to me.' The British officer said he was insulting. Crawford told him it could not work, and the Brit 'walked off in a huff'.

When you plug the incredible length of the BAMBI pipeline into modern pressure calculations they indicate that the 72 miles of pipeline was at the very edge of what was possible. Fuel could have moved through it, although we have to make some assumptions about how rough the internal pipe surface was and I am not competent enough to understand the rise and fall implications of the pipes. Any flow that does occur comes out as turbulent, not laminar and the pressure at the pumping end is, of course, the driving factor. We have assumed a length of the pipes at 72 miles, a pipe ID of 2 ⅞ inch (73mm) and a petrol density of 749 kg/m$^3$ and a dynamic viscosity of 0.00031.

There is also some debate also about the actual length of the pipelines laid over the channel. One report states that one ship expended 66.1 miles of HAIS cable in a geographic distance of 64.5 miles, but this cannot be correct as the actual point to point distance is 72 miles. In addition, the rise and fall of the pipes across the channel seabed must have added more distance to that. Indeed, one document report that the Conundrums were loaded with 80 miles of pipe.

The calculations do show that the pipeline could have worked, but at best each BAMBI pipe at 400 psi/28 BAR could have produced around 60 lt./min or 13 gallons/minute. At a planned 1,200 psi/ 83 BAR it could have produced 103 lt./min or around 22.6 gallons/min. Not a huge volume, but fuel none the less.

However, this calculation can take no account of the increased distances and back pressure due to the fall and rise of the pipeline across the channel (at 80 miles' flow rates drop again to just 12.5 gallons or 21 gallons per min) and the increased thickness of the petrol at the colder temperatures.

The figure for the DUMBO pipes, at just 22 miles are better. At 28 BAR each pipe could produce 114 litres/min or 30 gallons/min and at 83 BAR 195 litres/min or 51.5 gallon/min. For a while DUMBO operated 12 pipes - although as they continually failed they had to be replaced. Indeed, over just 22 miles the experience of the short DUMBO route was that the pipes struggled to survive for more than a few weeks at best.

Contemporary accounts say that with DUMBO the pressure used was 'around' 40 BAR to 'protect the pipes' and at this each pipe could move 136 litres/min or 30 gallons/min – 1,800 gallons an hour. Again contemporary accounts state that the DUMBO pipes actually achieved around 1,500 gallons per hour – the 27% error on my calculations could be due to the pipe internal roughness and the thicker fuel density at lower temperatures - but is also seems to suggest that the pipes had to be run at 30 BAR, when the calculation is within 4%.

(If anyone with a background in fuel flow in pipelines would like to further this work the authors would be pleased to hand over all their data! And publish their results!)

### So did any fuel flow ever flow though BAMBI?

In truth, we don't know. The evidence for it happening is circumspect or at best conflicting, the science against it is, as yet, incomplete. If it did work, it was a damned close run thing and the volumes possible, even at high pressure, would have been small. The pipe structure, its many joints and its huge length in an inhospitable environment were always against it.

## Was the Sandown pumping site ever completed?

Today all the evidence suggests that it was left incomplete, probably due to time and resources. Once it was realised that the Allies had achieved air superiority and that tankers could operator near the French shore it was not needed. Also, the American forces advanced rapidly so a complete backup site was not required. Sandown pumps were installed but it is now clear that the Sandown pumping site did not come on line at any time, and we suspect the pumps were never run or connected. We know that at least 8 diesel generators, unused and still wrapped in Cosmoline paper (a brown wax-like petroleum-based corrosion inhibitor literally smeared over anything metallic to protect it, or onto paper to form a rust proof wrapping) were taken from the Granite Fort (now the zoo) and dumped far out in Sandown Bay in 1945.

A huge amount of engineering effort and resources had been poured into BAMBI, but despite the best endeavours of everyone involved, it had only from one account, during its short period of operation, delivered just 3,300 tons (around 87,000 gallons) of fuel. This was equivalent to only one day of PLUTO pumping in the original plan and only enough fuel to keep the Allied armies supplied for less than a few hours. Compared with the total amount of fuel stored and used in the war effort, 3,300 tons was miniscule, and indeed some doubt has been cast on even this figure.

For example, by 1944, one storage depot constructed during the war in southern England, on its own, could store up to 180,000 tons and most 'Greyhound' tankers could carry 16,700 tons. The total amount of fuel delivered to the allied armies in North-West Europe from D-Day to 10th May 1945, when organised German resistance ceased, amounted to 5.2 million tons.

BAMBI, at best, had delivered less than 0.1 percent of this, but at a huge cost in resources.

The reality of the contribution of the undersea PLUTO pipelines to the success of the invasion of Normandy and Victory in Europe really relies on actually how much fuel was actually delivered by PLUTO undersea pipelines during the battle of Normandy.

The answer is none. It was not until 22nd September 1944, three and half months after D-Day that any petrol first flowed through PLUTO. By this time D-Day was over, Paris liberated and the allies had reached the Netherlands. Of course the PLUTO team were not responsible for the time taken to capture Cherbourg or indeed the rapidity and success of Patton's breakout.

Simply, the programme for getting the PLUTO pipelines operational was clearly unrealistic, and in places it seems that the planning process also failed. With the hindsight of 75 years, and understanding the incredible timescales and pressures the teams were under this observation may seem harsh.

However, some delays in PLUTO's timetable seem inexplicable and in part reflect a lack of planning to cope with possible differing outcomes. Prior to Cherbourg's capture, the Germans had systematically demolished all the port facilities such that it was not to be until late September that Cherbourg reached anything like full operational capacity. This was hardly surprising. General Eisenhower wrote that 'the thunder of the German demolitions in the port area reverberated from the surrounding hills.' The intention had always been to run the PLUTO lines into the harbour but, with that wrecked, over a month was wasted in debate.

Should the pipeline terminal now be located outside the breakwater? This would increase the difficulties with discharging the fuel, or should it be inside? This might endanger the harbour with possible fuel spillages, fires or explosions? Eventually it was decided to run the lines to a terminal sited in the bay of Urville-Nacqueville, outside of Cherbourg. Surely plans should have been put in place for such an eventuality long before D-Day?

There were also problems with the HAMEL pipelines as the Conundrums had to be overloaded to accommodate the fifty plus miles of extra pipe for the increased distance. While waiting for the liberation of Cherbourg, Conundrum1 had managed to gain ten tons of barnacles. They upset the balance of the Conundrum and the tides and rocky bottom on the sea at the Cotentin Peninsula were not favourable for pipe laying. However, it could be argued that all of this could and probably should have been anticipated by the PLUTO team.

One other factor has come to light. There is also a possibility that in the fog and chaos of war mistakes are made. When the pipe lengths arrived at the docks the process of welding them into one continuous pipe, some 80 miles long, began. (One source states 90 miles).

At intervals the sections were cleaned by compressed air being blown through followed by a piece of shaped wood being blown through at considerable speed. It appears from one eye witness account that at least some of these of cleaning woods were left in.

In 1946, while the pipes were being recovered, oxy-acetylene was being used to cut the steel pipe at the previously welded joints after it had been raised and brought on-board ship. At three of the joints the team were surprised to see that something caught fire as they cut the pipe. It turned out that the flames came from wooden 'bungs'.

They thought that these must have been driven in during the construction phase by somebody to render the pipeline useless. Officers were called and there was a visit from a special investigation team, who said they would be able to find out who was doing the work at the 'sabotaged' points on the pipe. I think sabotage is unlikely; it is more likely a manufacturing error. But at least one of the BAMBI PLUTO pipes could not have worked because it was blocked in three places down its length.

But what of TOMBOLA part of the PLUTO MINOR project? During the early phases of the invasion the allies had been pressing ahead with Operation TOMBOLA. On D-Day+19 the first ship to shore pipeline was laid to Port-en-Bessin which would allow tankers moored off the coast could discharge fuel. A second route was later established and five TOMBOLA lines were also laid to St. Honorine-des-Petres situated a few miles west of Port-en-Bessin to supply the Americans.

It would be another 18 days before the first fuel flowed through TOMBOLA on 3rd July 1944, (D+37), nearly a month after D-Day, but TOMBOLA proved to be only a limited success as the pipelines kept breaking. One of the causes for this was the rocky foreshore in the area where the pipelines were being laid. A small pipeline network was constructed from the two small ports of Port-en-Bessin and Ste. Honorine-des-Petres to a tank farm that was constructed at Mont Cauvin. The tanks had to be built above ground because of the very short time scale available, but they were heavily camouflaged to protect against enemy attack.

Using CHANT's (from Channel Tanker - a type of prefabricated coastal tanker) and other small tankers, up to 1,300 tons fuel per vessel was taken directly into Port-en-Bessin regardless of the risk of enemy attack. However, the weather and sea conditions were so poor that many of the prefabricated 'Chants' were too badly damaged for them to continue in operation.

By 28th July, with the stalemate in Normandy finally ending due to the success of the United States breakout codenamed Operation COBRA, no fewer than sixteen of the thirty-nine Chants were either being repaired or awaiting repair at Hamble in England, where a tanker repair facility had been constructed.

## *Was DUMBO a success?*

Those in charge of Operation PLUTO hoped that with the much shorter run from Dungeness to Boulogne, DUMBO would be more successful than BAMBI. The original plan had been that the DUMBO lines should be terminated on the beach at Ambleteuse near Boulogne, but this was found to be heavily mined. It was instead decided to utilise the outer harbour area of Boulogne even though this would make the approach for the cable-laying ships more difficult and that harbour area had also been mined.

The first run was made by HMS *Sancroft* on 10th October, but while the HAIS cable laying was carried out successfully great difficulty was encountered in securing the cable at Dungeness. With the weather deteriorating, it took until 27th October for pumping of petrol to start.

By mid-December a total of six HAIS cables had been laid (four 3 inch and two 2 inch), but only four of the cables were operational. The performance of these was also well below what had been expected of them. This was due to them operating at between 20 and 30 bar pressure instead of the planned 99 bar. As a result, daily deliveries averaged around 700 tons (18,500 gallons) instead of the planned 3,300 tons. (87,256 gallons)

According to the official history:

> *'There were frequent changes of plans and the enthusiasm of the PLUTO force gradually dwindled. In December 1944 the Royal Navy asked whether DUMBO like BAMBI should also be shutdown.'*

If that had happened, then presumably Operation PLUTO could never have been perceived as the success it generally was considered to be after the war. DUMBO was still well below its planned capacity when the Germans launched their final offensive in the West on 16th December in the Ardennes region of Belgium and Luxemburg which caught the Allies completely by surprise.

The target of that offensive was not the DUMBO terminal at Boulogne, but the port of Antwerp. On 2nd January, with the Battle of the Bulge largely over, a committee ruled that DUMBO should continue and that all the available HAIS cables left in store should be laid in an attempt to reach the originally planned throughput of 3,300 tons. DUMBO lines continued to be laid right up until the German surrender and even after the war in Europe had ended. The lines continued to be used up to the end of July 1945 to help supply the occupying forces, but were then closed down to free up the technical manpower required to operate them.

According to the Official PLUTO History, for DUMBO a total of ten HAIS cables and six HAMEL cables were laid of which eleven cables and pipes were operational giving, at peak, a flow-rate of 4,000 tons per day. However, according to Sir Donald Banks there were eleven HAIS cables laid. But the HAMEL pipes only had only a limited life expectancy before they failed. According to Donald Banks they lasted between 55 and 112 days, while according to the Official History they only had an average operational life of 56 days.

However, again according to the official history, DUMBO did not reach its peak flow-rate until after VE Day and that the average flow-rate was around 1,800 tonnes/day. This amount carried by PLUTO equated to 2 large tankers per month.

Major-General Christopher Tickell, CBE is a senior British Army officer who wrote that: *'We gained very little from PLUTO and DUMBO'.*

The Official History concluded that: *'DUMBO was more successful; but at a time when success was of less importance'.*

As the Official History also states that:

*'PLUTO contributed nothing to Allied supplies at the time that would have been most valuable – that is when no regular oil ports were available on the Continent and the Allies were relying on the unsatisfactory Port-en-Bessin.'*

## 2.5.1 PLUTO – A CONCLUSION

Today it is probably correct to state that the perception of those who have heard of the PLUTO project is that it was a great success and that the Normandy invasion would not have succeeded without it.

This propaganda started almost immediately after the end of the war in Europe when the secret of the PLUTO pipelines was first made public. Henley Cables quickly ran an advert referring to *'Operation PLUTO the petrol pipelines that made V.E. possible'*. Other companies soon joined in.

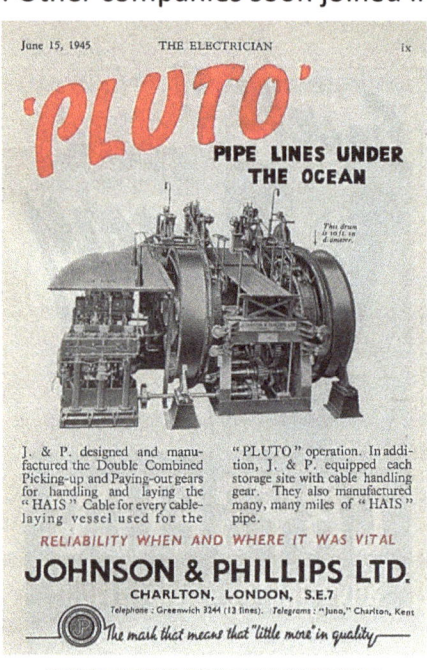

Captain Hutchings, Senior Naval Officer Commanding Force PLUTO wrote to all members of Force PLUTO advising them that they had *'contributed not a little to the final victory'* advising them that the Allied Supreme Commander General Eisenhower described it as *'Second only in daring to the Mulberry artificial harbours'*.

He also wrote that, *'This [PLUTO] provided our main supplies of fuel during the winter and spring campaigns.'* Reflecting after the war, Dwight Eisenhower stated that without the pipelines the U.S. forces would have run out of fuel. Today it is difficult to agree with these statements.

Lord Prior, in his foreword to Adrian Searle's book, wrote: 'that *Operation PLUTO was one of the outstanding engineering achievements of the Second World War is beyond question.*'

In part, I feel he is right, but Winston Churchill was nearer the mark when he declared: **Operation PLUTO is a wholly British achievement, and a piece of amphibious engineering skills of which we may be proud'.**

**Yes.** Without any doubt the PLUTO project was an epic achievement of planning, design, development, engineering, manufacture and installation. It wrote a whole new chapter in pipeline technologies against a background of total war, while under the most intense time pressure and competition for resources and manpower. Today, it is almost inconceivable that such a huge and complex project could have been achieved in such a tight timescale. In essence, this was less than a year, while the whole nation prepared to launch the largest amphibious landing in human history, against a determined and well dug-in enemy.

It is also a testament to the skills of the sailors under constant risk and fear of aerial or sea attack, that they tramped across the heavily mined channel at just 4 knots within two months of D-Day.

From September 1944, army engineers began rapidly building a pipeline south from Cherbourg that would eventually extend into Germany. It was, however, supplied by ocean tankers, not by PLUTO. A pipeline network was extended out from Boulogne into Germany. This network could be considered an extension of PLUTO, and petrol was pumped from the Mersey through PLUTO into Germany. The pipeline network that really did contribute greatly to the allied victory was not PLUTO but the unsung and largely unknown GPSS. It carried in excess of 250,000 tons per month of fuel through all of 1944 and the early part of 1945. The entire GPSS was an incredible endeavour for firstly keeping Britain in the war and then providing one of the means for victory to be won. With 20:20 hindsight, once the decision was made to carry out the invasion against the Normandy coast, then the concept of supplying the initial stages of the invasion by pipeline from England should have been abandoned.

Lookong objectively at the PLUTO project on the Isle of Wight, and taking into account the huge distances it had to cover, as engineers we are struck by several factors. The pipe bore of the HAIS and HAMEL pipes is too small for the volumes of fuel required and they are too fragile, both for the external sea pressure and the inhospitable environment of tides, currents and seabed. The internal pipe surface was too rough to allow smooth flow, the huge drop and rises across the deep channel would have caused a large back pressure with fuel already thickened due to low temperature. If this was not enough the pressure needed to force fuel all that way in any practical quantity, was simply too high for the thin pipes and newly welded joints.

Overall the Dungeness DUMBO lines were clearly more successful than BAMBI pipes from the Isle of Wight. The whole PLUTO systems at Sandown, Shanklin, Hungerberry and Thorness were simply abandoned within three days. Indeed, it now looks highly likely that the Sandown site was never even switched on or perhaps even tested. It is also possible that the pipeline feed to it around the town of Lake, and even the pumps were never completed. There are no surviving documents and of course no photographs that show the layout of the pipes or the ring main at Sandown, if it was ever installed. One eye witness account remembers that a large number of generators from the Granite Fort were still wrapped in their protective covering and had never been used, when they were dumped out at sea after the war.

By the end of the war the entire PLUTO project, both from Dungeness and the Isle of Wight only delivered 370,000 tons, (8 per cent) of the total fuel supplied to the Allied forces on the Continent. The rest was delivered by tanker, either in bulk or in cans, or indeed by airlift. Sir Donald Banks' book increases the 'PLUTO' total to 575,000 tons by including fuel delivered after the surrender of the German forces.

Perhaps the one success that came out of PLUTO on the Isle of Wight was the SOLO pipelines over to the Thorness beach manifold. From there a main pipe ran around to Rew Street Point and a small buildings that was essentially just a valve rooms. From here two pipes ran out under the sea, at the end of which were floating pipes to refuel ships – not for their engines (which ran on heavy oil - essentially marine diesel) but the tanks of the CHANTS and other tankers heading for the beaches of France. Recent research and analysis of modern aerial photographs has shown a second pipeline heading out to sea lies just beyond the collapsed Rew street gun battery remains on the beach. There are parts of a concrete structure the same size as the Rew Street building and a single pipeline can be seen on the sea bed. There is even a possibility of another set of pipelines further round the headland near Gurnard – more research is needed! The Rew Street refuelling pipes gave the navy much needed additional capacity and refuelling areas well away from main ports and dockyards. This was critical due to the risk of air attack or accident, while thousands of gallons of highly flammable petrol were being moved.

One outcome of the PLUTO project was that the pipelines developed in 1944 were the forerunners of all flexible pipes used in the development of today's offshore oil fields. Indeed, during and after the Falklands conflict, 10 inch diameter floating pipelines fed into shoreside tanks. From there a pump running at 4,000 psi pushed fuel up a 6.5 mile (10.5 km) high pressure pipeline, that climbed over 1,600 feet. A direct descendant from the 1944 project - PLUTO had gone to war again.

The end of World War Two did not see the end of the secret GPSS pipeline (even though the Isle of Wight and most underwater sections were immediately recovered), in fact it was just the beginning. Today the GPSS pipeline system that fed PLUTO has been extensively extended and updated, certain sections have been renewed, relaid and diverted until it now covers approximately 2,500 km of pipe and associated storage depots, pumping stations and other sites that now supply most of this countries' modern airports.

But in 1944 the PLUTO project must be considered a failure, as the critical fighting in Normandy was over before any PLUTO fuel began to arrive. But there are some other factors that can be considered. It can be argued that the very possibility of PLUTO gave the Allied command the confidence to attack the Normandy beaches. Given that tankers were not set ablaze and destroyed by the Luftwaffe during the invasion, another defence of the cost and resources that PLUTO consumed during WW2 can also be that 'it saved a very large tanker tonnage which was badly needed in the East.'

In the end I think it is fair to say that PLUTO on the Isle of Wight was a probably a Pipeline too far. The technology of pipe manufacture and laying that had been rushed so quickly into operational combat use was not ready and could not hope to survive the rigours of 72 miles (plus) of English Channel storms, tides, shipping, currents and geology. As it was, the 22 miles of pipelines across the Dover Straits had to be continually replaced, with the life expectancy of any one pipe often just a matter of weeks before it failed. The technology to recover and repair had not been developed, so they were simply discarded and a new one laid.

But overall, the fact that the PLUTO project on the Isle of Wight had a limited impact on the outcome of the war does in no way detract from the importance of this massive, innovative and secret project. Alongside the Mulberry harbours that were constructed immediately after D-Day, Operation PLUTO is considered to be one of history's greatest feats of military engineering.

But above all else it is a testament to skills, dedication and fortitude of the men and women who developed and built it, and the sailors who laid the pipelines. It was without doubt one the most audacious and complex engineering projects of WW2.

PLUTO ceased all operations in July 1945, just a few months after the end of the fighting in Europe.

**The team behind PLUTO disbanded on 31st August 1945.**

## 2.5.2 PLUTO AFTER THE WAR

After the war the PLUTO pipelines installed at such considerable cost were no longer secret and various parties moved to publicise their 'success'.

However, the PLUTO pipelines were not only no longer required, they had become a nuisance. Ships repeatedly fouled their anchorages on the HAIS pipes and the Royal Navy cut and removed sections that were just a few miles offshore. In addition, there was significant salvage value (and requirement for post war rebuilding) in the lead in the HAIS cables and the steel in the HAMEL pipes. Starting in August 1946, the former HMS *Latimer* and HMS *Holdfast* were transferred to the Ministry of War Transport, (renamed Empire *Ridley* and Empire *Taw*) and were used along with other vessels in a large scale recovery project. The first part of the operation used a grapple to find a pipeline and haul it up and onto the ship's bow. (Which was some task, bearing in mind that 100 miles of 3-inch pipe weighed 6,000 tons.) The HAIS pipe was found to be in good condition, and its high lead content made salvage particularly valuable in a country trying to rebuild after 5 years of war. Lengths of HAIS pipes were cleared of petrol and cut just once before being coiled in the ship's hold. Once ashore it was cut into lengths suitable for transportation by rail and sent to Swansea where the recovered lead was melted and cast into ingots; the wires were straightened and used as rebars; the steel tapes were flattened and used to make corner reinforcements for heavy duty cardboard boxes; and the jute was made into blocks that could be burned as fuel in a furnace.

The HAMEL pipes were also valuable, but being less flexible, it needed to be cut into lengths on the deck of the recovery ship. Cutting either type of pipeline was very dangerous because the pipes still contained petrol. The contaminated petrol from both types of pipe was recovered and cleaned up, yielding some 66,000 imperial gallons of useful fuel. The salvage operation lasted three years. Of the 25,000 tons of lead originally used in the HAIS pipelines, 22,000 tons were eventually retrieved, at a time when the scrap value of lead was £55 per ton and it was in great demand. The reclamation of the HAMEL pipes was less successful with 3,500 out of 5,500 tons of steel being recovered.

*PLUTO salvage operations ceased in July 1949.*

# Where PLUTO Crossed the Path

*Rambles with a Purpose*

*on the Isle of Wight*

APPENDIX 1

SOME NOTES ON WALKING ON THE ISLAND

(IN GENERAL)

(This may well ramble a bit, but then that's what it's all about!)

Whilst I have given these walks some purpose in themselves, do take the opportunity to encompass any adjacent points of history or other interest; especially if the time of your stay in the area is limited.

In this appendix I offer my views on a variety of walking associated subjects which I hope the reader will find of use or interest although no doubt experienced walkers will have their own views.

### *HISTORICAL ASPECTS*

There is no better way to absorb the history of an area than by a visit to the church and churchyard and in some cases the over-spill cemetery. Unfortunately, due to the age in which we live many churches are now locked. Access can usually be obtained but this requires some forward planning which is hardly practical when casually passing by.

However, it's always worth a try, some are still open and churchyards are rarely locked. As Fred Hollis recounts in his book 'To Heaven in a Glass Wheelbarrow' when the Parson asked a man if he would like to contribute to some new gates for Brighstone cemetery the man replied:-

'I don't know if I shall zir,

I can't zee as what they wants a gate on there at all fore,

nobodys is in a hurry to get in there

and they what be in there baint gwine to breakout'

There is a wealth of stories in tombstones, the families of the area, the tragedies, and of course the occasional, distinctive, 'war grave' with its regulation Imperial War Graves Commission headstone.

Notice also how many members of one family are listed on the War Memorials, especially in the Great War of 1914-18, and often some of their next generation in the 1939-45 war.

Also noticeable on the Island (indeed throughout Britain) are the number of eldest sons, of leading families, who were lost whilst leading their men into the slaughter of that First World War. In many cases these were clever well-educated men who would have become local and national leaders; a waste that we still suffer from! The detailed memorials to these particular young men tend to be inside the churches.

I have a personal shudder when visiting Arreton Churchyard as there is an old gravestone near the main door, just legible if the light is right, inscribed 'John Farthing of Appleford 1647'! (Not an ancestor, as far as I know)

The Island is particularly rich in ancient manors and other interesting houses. These have been much written about elsewhere and I would recommend three relatively modern books on the subject:-

'The Manor Houses of the Isle of Wight' by C W R Winter, (1981)

'Castles to Cottages' by Johanna Jones. (1999) Both were published by the Dovecote Press.

'Farmhouses and Cottages of the Isle of Wight' compiled by M. Brinton

Published by The IOW County Council (1987)

In addition, I have listed in the bibliography a selection of other books about the Island; of which there are many.

## *DIVERSIONS OF FOOTPATHS*

This is an issue on which I have very strong views. That is on the legal provisions for the diversion, or occasionally closure, of footpaths. I have been walking the footpaths on the Isle of Wight for over fifty years and I am continually infuriated by unacceptable diversions.

They are mostly all done quite legally and above board, (I have reservations about some early ones) but are nearly always done at the request of, and in the interest of the resident or landowner. It would be most unusual if a diversion were requested to suit the interested walker; indeed, if the footpaths had been left as they had evolved the walker would probably be very happy with their routing.

Despite objections, by the Ramblers Association and others, it is very rare for a deviation request to be refused, even when it goes to enquiry.

The most infuriating are those diversions where the path originally went near a manor, building or perhaps through a farm and have been diverted around it. Now this may appear to be nitpicking, but many of these places are of historical, social and architectural interest; indeed some have details mentioned in the older guide books. A classic example is Great East Standen manor but there are many more; some recent ones still causing controversy.

One particularly galling point is that if an objection is made it is necessary to give a reason. However, it would appear that no reason has to be given, at least for public perusal, by those requesting the diversion. If a reason is given there seems to be no requirement to publish it in the legal order!

My view is, that the duty of the authorities should be to maintain the status quo except in exceptional or impossible situations; (eg cliff falls). Assuming that legal searches were competently done, land owners and residents bought their properties knowing that the right of way existed; and there the matter should have rested.

Whilst the authorities appear to be too accommodating to applicants, it seems that they may have little option. I suspect that the legal provisions are at fault and have been instrumental in taking away a lot of what had previously been of interest from our footpath network.

## *ACCESS TO THE FOOTPATH NETWORK*

Despite my concerns about diversions some paths have been added and there are now some 'permissive' paths. So the Island has an enviable wealth of footpaths and bridleways. I have never set out to walk all of them, so even after half a century I can still find parts of paths that I haven't walked before. I said in my preamble that it is much better to look for PLUTO markers in the winter. The bare branches and hedgerows allow much more visibility both in panorama and in detail. However, don't confine these walks to the winter as all the seasons have their attractions, especially with regard to botany, birds, and butterflies.

Time seems to be a problem to us all, yes even to us OAP's, so many tend to use a car to get to the starting point of a walk, a bus is a better option, albeit more expensive for some; often a combination of both is useful. The bicycle is another possibility providing you can be happy about leaving it. If the 'walk' is on byways and or bridleways you could of course use the bike.

It is always nice to make the walk circular; in some cases I have suggested how this could be done but it is not always practical, especially as some stretches of road are so dangerous to walk along. It is in these cases that a bus is most useful, with or without the car in combination.

A major problem with the use of a car, in getting to the start of a walk, is the problem of leaving it; in regard to both security and road safety. I have occasionally suggested where a car could be left but it is obviously up to drivers to satisfy themselves in this. Where, in my opinion there is nowhere to leave a car I have either said so, or ignored the possibility. Cars parked in out of the way places are vulnerable to thieves so the usual precautions should be taken. A place in full view of the road would be preferable.

Also worth a warning is that parking on the verge, even though well clear of the road, is not always entirely legal; there are some, to me, complex laws depending on the classification of the road ie. whether it is a highway, carriageway, thoroughfare etc. Personally I have never had a problem and felt happy as long as I have been well off the road and not making a mess of the verge. It would be nice to think that the authorities would take a sensible view on this, but don't bank on it. (No pun intended).

For many of us getting out of town in the first place is the problem, this can be boring, hence the use of cars.

## *HAZARDS WHEN WALKING*

I don't intend to bore you with such things as apparel and footwear; I am sure you will walk in whatever you feel comfortable with. Suffice perhaps, just to say that flip-flops are definitely out and on a closely allied subject do not take the possibility of adders too lightly; we have plenty on the Island and some are well above average size. So watch where you're treading, especially on the downs and when poking around in hedgerows.

Bulls and dogs are other subjects worth mentioning; there are some pretty irresponsible people around.

With respect to bulls in fields crossed by footpaths, it is worth being cautious. Do not assume that it would not be there if it was a danger; the farmer may think it's alright but many a farmer has come to grief on that basis. Unfortunately, the law in this respect is complex; in certain cases a farmer can quite legally put a bull in a field crossed by a footpath, it depends on what animals are in the field with him and on his breed. I am not sure if this law has been amended with respect to recently imported breeds.

One has to weigh up the situation oneself; this may include keeping close to an easily negotiable boundary rather than following a path through the centre of a field (quite justified, in my opinion). If in doubt retreat, certainly treat all bulls as potentially dangerous. Also bear in mind that a cow separated from her calf can be a problem; often this will be the one making a lot of noise probably near farm buildings; give her a very wide berth away from the buildings.

For the law about bulls on Rights of Way see The Wildlife and Countryside Act 1981 Section 59 (page 288).

Section 2 of The Animals Act 1971 (page 228) and The Health and Safety at Work Act 1974 (page 236).

The Animals Act 1971 Sect.2 (page 220) also covers dangerous horses on Public Rights of Way.

There may be more recent amendments. (We have more 'foreign' breeds in the countryside now!)

Dogs are an entirely different matter, in a public place they should be on a lead but this is often impractical especially with working dogs and some owners have trouble holding large dogs even if they are. We have all heard the cry 'he's OK' and usually 'he' is but it's the dog who thinks he's guarding his patch that has to be looked out for, especially when passing a farm or other buildings. If you are on the path and the dog has his hackles up and teeth bared you have every right to defend yourself.

I am a great believer in the old adage 'speak softly but carry a big stick' but it's a big mistake to hold the stick aloft in a threat to strike; certainly hold it between you and the dog as a barrier. Do not look a big dog in the eye, that's a challenge, talk to the dog softly but firmly and back off slowly, don't run; if necessary use the stick by pushing it out in front of you ie. not raising it. If it presses home an attack, I can do no better than to quote from 'Richards Bicycle Book' which some may not be happy with. 'Be confident' he says 'make sure the dog realises that you are big, dangerous, and a serious contender in a fight' give him something to bite on, your bag or coat and only use the stick as a last resort!

If bitten you need to get medical advice, but even if you escape unscathed you must inform the police and if practical the owner of the dog; a child coming along later may not be able to withstand an attack.

However, in over fifty years of walking on the Island, I have only twice had a real problem and on both occasions the result has been a draw; slightly in the dogs' favour as in both cases as I was deviated from my intended path. (OK so the dog won). That is not including the occasion when a Jack Russell got such a firm grip on my trouser leg, whilst I was cycling past Billingham, that he went round with the crank for about twenty yards, before I could shake him off.

## GATES

The Country Code use to say 'close all gates' (although some versions of the code do now say 'leave gates as you find them.') I only believe this to be correct in certain situations e.g. If you yourself have opened a gate or where, if found open, stock are obviously going to escape onto a road.

There is nothing more infuriating, when moving stock either back along a road after milking or from one field to another, than to find that some helpful soul has closed the gate that you opened half an hour ago. So, generally, unless it's obviously wrong, gates should be left as you find them.

Another small but important point, if you have to climb over a gate, please do so at the hinge end; it's kinder to the gate and usually more stable.

## WALKING ON ROADS

Many of the Island's roads are extremely dangerous to walk along; obviously not those with pavements. The code tells us to walk facing oncoming traffic! All else being equal, that is probably a sensible practice; unfortunately, things are rarely equal and there are grave dangers in this practice if followed slavishly.

I would contend that it is far more important to walk where you can be seen irrespective of which side of the road that happens to be. That 'boy racer' will be taking a racing line through that blind corner with nearside wheels in the gutter, quite oblivious, like the majority of motorists, of the fact that there could actually be someone walking on the unseen side!

No, it's much more important to walk where you can be seen, even if it means crossing the road a few times; in a series of bends. And don't be too proud to jump onto the verge or even into the hedge if necessary.

# Where PLUTO Crossed the Path

*Rambles with a Purpose*

*on the Isle of Wight*

## APPENDIX 2

### 2.1 BIBLIOGRAPHY and FURTHER READING

Adrian Searle's book on PLUTO covers the subject well, while Tim Whittle's book *'Fuelling the Wars'* on the Government Petroleum Supply System (GPSS) tells the complete story of the network that gave birth to PLUTO and was ultimately much more important to the conduct and success of the entire Second World War. Sir Donald Bank's book is a key reference source as after all, he was there although concerns over what the censor cut and amended must be considered. *Fuel to the Troops* gives us the only eye witness account of PLUTO in Ryde on the Isle of Wight and answered just a few of the puzzling questions.

**PLUTO PipeLine Under The Ocean - The definitive story by Adrian Searle** Publisher: Shanklin Chine; ISBN-13: 978-0952587606

**Tim Whittle - Fuelling the Wars: PLUTO and the Secret Pipeline Network 1936-2015** Folly Books Ltd (28 April 2017); ISBN-13: 978-0992855468

**Sir Donald Banks - Flame over Britain: A Personal Narrative of Petroleum Warfare**, London 1946. Sampson Low (1946); ASIN: B0007IZLKE

**Fuel to the Troops: A Memoir of the 698th Engineer Petroleum Distribution Company 1943-1945. John Sullivan.** Publisher: CreateSpace Independent; ISBN-13: 978-1470028732

**PLUTO - World War II's Best-Kept Secret by Bob Knight, Harry Smith & Barry Barnett.** Published in 1998 by Bexley Council. Softback, 34 pages with many illustrations about the involvement of The Callender Cable Co.

The internet now hosts a number of fascinating sites on PLUTO including the Combined Ops site https://www.combinedops.com/ and a study of the Hungerberry Woods site http://www.iwhistory.org.uk/RM/copse/fuel.htm.

A recent study of the Shanklin pumping sites has also significantly contributed to the story of PLUTO on the Isle of Wight; http://www.iwhistory.org.uk/RM/shpumps/home.htm .

# 2.2 SELECTED ISLAND BIBLIOGRAPHY

There are now so many books about the Isle of Wight that I could not possibly list even all those which I have found interesting and informative. Therefore, I will confine myself to some of those which people may find helpful in filling out the snippets that I have included in this book. I have already mentioned those which particularly pertain to my subjects; these are some others; they are in no special order but just as they come to mind.

**'A Short Account of the Geology of the Isle of Wight'** by H.J.Osborne-White. HMSO 1921(1975)

**'Wight: Biography of an Island'** by Paul Hyland. Victor Gollanez Ltd. 1984

**'Newport Isle of Wight Remembered'** by Bill Shepard. Isle of Wight Natural History and Archaeological Society 1984

**'Coastguards of the Isle of Wight'** by Tony Gale. Coachhouse Publications 2005

**'Isle of Wight Within Living Memory'** I of W Fed. of Woman's Institutes Countryside Books and IOWFWI 1994

**'Isle of Wight at War'** by Adrian Searle. The Dovecote Press 1989

**'The Royal Prisoner'** by Jack Jones. Lutterworth Press 1965 (1978)

**'The Isle of Wight an Illustrated History'** by Jack and Johanna Jones. The Dovecote Press 1987

**'The Enchanted Isle'** by C. W. R. Winter Cross Publications 1990

**'The Countryside of the Isle of Wight'** by John Wolfenden. Crossprint 1990

**'Historic Parks and Gardens of the Isle of Wight'** by Vicky Basford. Isle of Wight County Council 1989

**'England's Garden Island'** by Ethel C. Hargrove. Isle of Wight County Press 1926

**'The Isle of Wight'** by Davenport Adams. Nelson and Co. 1884

**'The Isle of Wight'** by Edmund Venables. 1860

**'A Driving Tour of the Isle of Wight'** by Herbert Garle. The Isle of Wight County Press 1905

**'Marks Corner, Past and Present'** Compilation
- Marks Corner Millenium Group 2001

**The Place Names of the Isle of Wight'** by A. D. Mills. Paul Watkins of Stamford

**'Gurnard, a Village and its Church'** by Sheila Caws. All Saints Church Council 1992

**'One Hundred Years Before the Mast'** by Brian Greening.

**'From Sea to Air'** by A. E. Tagg and R .L. Wheeler. Crossprint 1989

**'Love of an Island'** by Richard J. Hutchings. Isle of Wight County Press 1989

**'Rookley'** Compiled by The Rookley Parish Book Committee 2002

**'The Mills of the Isle of Wight'** by Kenneth J. Major . Charles Skilton Ltd. 1970

**'The Storey of Victorian Shanklin'** (1977) and **'Shanklin Between the Wars'** (1986) both by Alan Parker. Peter Spiegl & Co.

**'The Cartulary of Carisbrooke Priory** (1981) and **'The Charters of Quarr Abbey** (1991) both compiled by Dom S. F. Hockey.  The Isle of Wight County Record Office.

**'Insula Vecta'** by  Dom S. F. Hockey.   PhIlimore & Co. Ltd. 1982

**'Domesday Book,  A Complete Translation'**   (HMSO) Penguin Classics 1992

(A thought to leave you with; HMSO is now **owned** by a German company!)

*One of the Rushton generating engines in the PLUTO Pavilion now under restoration, The volunteer team are hopeful it may well run again!*

# Where PLUTO Crossed the Path

*Rambles with a Purpose*

*on the Isle of Wight*

## APPENDIX 3

### 3.1 OTHER PLACES OF PLUTO INTEREST

After the end of the war, most of the PLUTO pumps were sold for scrap and the diesel generators were reused or indeed one eyewitness remembers them being dumped far out in Sandown Bay. Until recently only two pumps were known to still exist - one at the Imperial War Museum, Duxford and one at Bembridge Heritage Centre on the Isle of Wight.

If you would like to know more about the wartime history of this area of the Isle of Wight we recommend that you visit Bembridge Heritage Centre. Their displays cover a wide range of historical interest and new displays are introduced each year to commemorate special events or topics. Scale models, for example Bembridge Railway Station (closed in 1953), are a special feature of the Centre, and a fully-restored PLUTO pump (Pipe Line Under the Ocean) is on display with the story of the PLUTO project. The Centre is located at the rear of the library in Church Road, Bembridge. Opening times: April to October, Monday, Wednesday and Friday 11:00 am -3:00 pm and Saturdays 11:00 am - 1:00 pm.

We now know that a third pump survived. Until 1997 it was used to wash insulators on England's tallest electricity pylon by the River Thames. In 2007 the pump was offered by the National Grid for restoration, and thanks to the efforts of Robin Maconchy at Bembridge Heritage Centre, it was brought home to Sandown Granite Fort (now the IOW Zoo in Yaverland Road, Sandown, PO36 8QB. Open daily 10-5). Now, after careful restoration, the pump can be seen again in its original setting along with a good exhibition.

The Shanklin Chine Heritage Centre (Shanklin Chine, 3 Chine Hill, Shanklin, Isle of Wight PO37 6BW) (10.30 -7.30 - Seasonal – please check website) offers a good presentation and artefacts associated with PLUTO. The chine also displays a length of pipeline, although it diameter is not a standard PLUTO size.

The recent full relisting by Historic England of the PLUTO Power Pavilion at Sandown on Brown's golf course brought together a group of experts from diverse historic and technical backgrounds. This generated a large amount of research, some of which is presented in this book and in the preserved building on the edge of Brown's Golf course. You can of course visit the Brown's golf course café and take tea **exactly where a PLUTO pump once sat!**

*In 2017, Tim Wander gave his first (standing room only) PLUTO lecture at Brown's Golf Club in the PLUTO pump room. From this the Smallbrook tanks and the Ryde TOMBOLA system were discovered, the PLUTO Pavilion restoration began, and this book was born.*

*Beach flame barrage test - were systems installed to protect Shanklin and Sandown?*

## 3.2 MORE PLUTO QUESTIONS TO ANSWER!

It seems that very little of the PLUTO project on the Isle of Wight was documented. This was possibly because of the huge rush and almost impossible timescale that the project was conceived and executed in. Another reason may be that it ran for such a short period of time. Of course PLUTO research is also not helped by the fact that after 75 years or more not many people remember it - although some witnesses have come forward and new clues come to light seemingly every day, in part driven by a series of PLUTO lectures that have been very well attended across the Island.

This new information, and many walks have allowed us to complete the definitive route for the PLUTO pipeline across the Island and for the first time document the Smallbrook refuelling tanks and TOMBOLA pipeline work in Ryde. It has also started the research on the Rew Street refuelling site – but more information is needed. The Smallbrook site is intriguing. The visible scar that shows on the 1945 picture is far larger than a C2 tank – we now suspect this may have been another large storage tank similar to Hungerberry. But if it was (or even if it was several C2 tanks), how were they filled?

But their presence would solve one huge problem that has never been addressed with PLUTO at Shanklin. That is what happened to all the fuel trapped on the Isle of Wight when BAMBI abruptly shut down. There is no record of hundreds of tankers taking it away. It couldn't be pumped back to Thorness, and indeed the back pressure issues of this pipeline and the tank could have been catastrophic.

Assuming the Hungerberry tank was filled – (even though no eye witness reports can be found of people even smelling petrol (it was open to the sky)), when the PLUTO/BAMBI system was switched off, presumably it was full of fuel - some 620,000 gallons.

Add to this the huge amount of fuel 'trapped' in the main 8-inch pipe over 14 miles across the Island. (Around 161,000 gallons). Then add in the 8-inch Lake to Sandown feed pipe 5 miles (8km) holding another 57,000 gallons. Even the two HAMEL 3-inch pipes under Sandown Bay, 4 miles each, would have contained 12,000 gallons of petrol. Of course once the pipes across to Cherbourg burst, all that fuel was lost into the sea until the pumps could be shut down – we estimate at least 213,351 gallons was lost, probably much more.

So what happened to all this Island petrol? We have a theory.

It would also align several vague references of a PLUTO pipeline from Ryde to Shanklin, possibly more tanks in the Whitefield woods and it could explain whats happened Smallbrook.

We think there may well have been a PLUTO pipe from the Sandown site, probably alongside the railway line into Ryde. The Smallbrook tanks would have been a useful store for filling ships away from the railway and the risk of explosion or fire. It's only around another 3.5 miles from the Granite Fort, and around 1.5 miles from the Smallbrook tanks to Ryde. The pipeline would have been lost alongside the track, easy to lay on the surface, and would have attracted no interest being a non-pedestrian access area.

So when the BAMBI pipelines failed/were shut down was Hungerberry tank emptied via the Smallbrook tank(s) and out across the Ryde beach to waiting tankers? – In fact did PLUTO on the Isle of Wight actually fulfil some of its promise?

In the same vein we can assume that the Badminston storage tanks near the Fawley refinery on the mainland were emptied across the Solent to the Thorness terminal, where the Rew Street and possibly Gurnard TOMBOLA pipes would have loaded tankers and moved fuel to Northern France.

One day we will find out.

There is an ongoing frustration with researching PLUTO on the Isle of Wight. In reality, there are only perhaps five or six photographs that we can conclusively state show PLUTO operations on the Isle of Wight. There are no smiling groups of engineers, no diagrams, schematics, pump layouts, wiring diagrams or engineers notes. We do not know how the pumps at Sandown (and in truth Shanklin) were connected. Or even if they were connected?

In our many discussions John and I have even discussed the possibility that the entire PLUTO project on the Isle of Wight was mothballed before June 1944, and perhaps the pipeline never carried any fuel? Perhaps Hungerberry was never filled, the pumps were never run, and as the war moved rapidly east a cover story was put out to save face. Post-war, every trace was quickly removed. We just don't know. But the evidence is mounting.

Of course with all the new research and information come other clues, tantalising snippets of information that demand further research and this book is part of that on-going story. If you know of any new information, new photographs about project PLUTO please contact the authors via the website or at one of our lectures.

We are especially looking for information on the Sandown Defence Plan (SDP) - no copy of it can be found – possibly because it authorised the deployment of poisonous mustard gas which if used would have been in direct contravention of the 1925 Geneva Convention. But Churchill gave the order anyway, and many RAF bases and beaches were 'protected' by gas across the country.

Beach Flame Barrage systems were also deployed across the south coast of England. The plan was that these systems would flood the beach with ignited fuel to literally incinerate an invading army, as they stormed the beaches.

References have been found to a beach flame system at Freshwater, where the tanks were gravity fed from the hill above. We now believe that the same system was deployed at Sandown, possibly the closed and dug up Bembridge airfield (although this was more likely defended with poison gas) and perhaps even Shanklin, although the proximity of the town above the cliff would have been an 'issue'.

This was backed up by a series of horrific flame 'fougasse' traps – essentially crude but enormous 'Napalm' bombs, designed to be a last ditch defensive position.

Other questions remain. Did the BAMBI pipeline ever pump fuel or did it fail completely during installation? If it did work, why did it fail so quickly, and why was it abandoned immediately?

Just how were the two PLUTO pipelines across the channel to Cherbourg connected, and from where were they actually fed?

We now think that the entire Sandown pumping station and its pumps were never run, and probably never completed, but how much work was done?

*There is still more to tell about the story of PLUTO on the Isle of Wight.*

*John Farthing    Tim Wander.*

*July 2019.*

# Where PLUTO Crossed the Path

### *Rambles with a Purpose*

### *on the Isle of Wight*

**JOHN FARTHING** is a former Chartered Engineer who, before retirement, was employed at the Cowes Radar establishment as a Project Manager. He came to the Island in 1953 for an apprenticeship with the, then, Saunders-Roe aircraft company.

Being a countryman at heart he soon discovered the delights of the Island's countryside and has spent the last sixty-five years enjoying it, especially its footpaths and bridleways. A four year (part time) spell in training as an RMFVR Commando Troop Signaller was an additional 'walking' experience.

Despite having travelled widely in his work, John says **he has never had any desire to live anywhere other than the Isle of Wight.**

**TIM WANDER** is a former Chartered Engineer who tried to retire a few years ago - but somehow managed to get even busier. Today he is an historic consultant, author, lecturer and a specialist in historic building renovation. Raised and educated in Melton Mowbray in Leicestershire, Tim spent the first 17 years of his career with the Marconi Company, working across the world. In 1999 he then started Project Managing a series of specialist building projects across the world.

Tim has written many books about the early days of radio broadcasting, Marconi and three radio plays. His first stage play debuted at Northwood House in November 2018, two more are in production, and his first film script is well underway. Tim has been part of the prestigious Isle of Wight Literary Festival for three of the past four years – in 2018 to a capacity audience of over 165. In 2019 both Tim and John will be back there with their PLUTO story. Tim regularly acts as a technical and historic consultant and provides interviews and articles on the early history of radio and broadcasting. In 2016 Tim took over as Consultant and Curator for Science and Industry for Chelmsford City's museum service. He is now developing plans to celebrate the various centenaries of British radio broadcasting in 2020 and 2022.

For nearly thirty years Tim has been fascinated with the military and social history of the Isle of Wight and in 2018 he published *Culver Cliff and the Isle of Wight at War.* But it was the first edition of *Where PLUTO Crossed the Path* that first ignited Tim's interest in the PLUTO project photocopied from a rare (and now lost) example in Sandown Library. This led him to walk the PLUTO route on many occasions and drove his involvement with the ongoing restoration and research surrounding the PLUTO pavilion on Sandown golf course.

Tim lectures all over the UK (and on cruise ships) on the history and career of Marconi, early wireless and broadcasting, military history, PLUTO and the amazing history of the Isle of Wight. His informal and fully illustrated lectures are always driven by his enthusiasm and love of history, aided by an almost encyclopedic knowledge of his subject.

## *Other books by Tim Wander*

**2MT Writtle - The Birth of British Broadcasting (1988)**

**Marconi on the Isle of Wight (2000)**

**2MT Writtle - The Birth of British Broadcasting (2010)**

**MARCONI'S NEW STREET WORKS
1912 – 2012**

**Birthplace of the Wireless Age (2011)**

**Marconi on the Isle of Wight (2013)**

**The Marconi Company and Writtle (2013)**

**Northwood House. A Guidebook, History and Tour (2015)**

**Guglielmo Marconi - Building the Wireless Age (2015)**

**MARCONI'S HALL STREET WORKS
1898 – 1912**

**The World's First Wireless Factory (2016)**

**The Ghosts of Northwood House (2018)**

**Culver Cliff and the Isle of Wight at War (2018)**

**For more information on these books and new projects please see:
www.marconibooks.co.uk**

*PLUTO Markers ............Walk 1 and 2*

*PLUTO Markers ...............Walk 3 and 4*

*PLUTO Markers ...............Walk 5 and 7*

*Pluto Markers ..................Walk 9 and 10*

*PLUTO Markers ...............Walk 11 and 13*

*PLUTO Markers ...............Walk 14 and 15*

*PLUTO Markers. Walk 16, 19 and 20*

*But did you find this one?*

THE END

www.ingramcontent.com/pod-product-compliance
Lightning Source LLC
Chambersburg PA
CBHW040519220526
45473CB00013B/2922